Cosmos and Revelation

Cosmos and Revelation

Reimagining God's Creation in the Age of Science

Peter R. Stork

WIPF & STOCK · Eugene, Oregon

COSMOS AND REVELATION
Reimagining God's Creation in the Age of Science

Copyright © 2021 Peter R. Stork. All rights reserved. Except for brief quotations in critical publications or reviews, no part of this book may be reproduced in any manner without prior written permission from the publisher. Write: Permissions, Wipf and Stock Publishers, 199 W. 8th Ave., Suite 3, Eugene, OR 97401.

Wipf & Stock
An Imprint of Wipf and Stock Publishers
199 W. 8th Ave., Suite 3
Eugene, OR 97401

www.wipfandstock.com

PAPERBACK ISBN: 978-1-6667-3027-2
HARDCOVER ISBN: 978-1-6667-2154-6
EBOOK ISBN: 978-1-6667-2155-3

OCTOBER 11, 2021

In the embrace of *Abba*

To Elsbeth of blessed memory,

for fifty-eight years of loyal love.

To Barbara, for golden years.

He who descended is the very one who ascended higher than all the heavens, in order to fill the whole universe.
EPH 4:10 NIV

The truth of our faith becomes a matter of ridicule ... if any Christian, not gifted with the necessary scientific learning, preaches as dogma what scientific scrutiny shows to be false.
—THOMAS AQUINAS

Contents

Acknowledgements | xiii
List of Illustrations | xv

Introduction | 1

An Autobiographical Note | 1
How the Book Came to Be Written | 4
Obstacles Ahead | 9
The Challenge for the Churches | 11
Affirmations, Limits, Advocacy | 14
Conclusion | 16

∾ 1 ∾
The Nature of Revelation | 17

The Dimension of Mystery | 17
A Brief Theology of Revelation | 20
Revelation: The Cosmic Context | 25
Revelation in History | 28
Personal and Public Revelation | 31
The Problem of God's Hiddenness | 34
Revelation or Science? | 35
Science as Revelation? | 38
Conclusion | 41

∽ 2 ∾
The Biblical Account of Creation | 43

Text and Revelation | 44
Approaching the Text | 46
Inspiration | 48
Names for God | 51
Composition of the Text | 52
More Issues of Interpretation | 53
"In the Beginning . . ." | 53
The God Who Creates | 54
Elohim Speaks | 55
Eretz (Heb. Land, Earth) | 56
"Here Is the Good" | 57
"Let the Earth Bring Forth . . ." | 58
Time | 59
"According to Their Kind" | 60
God's Dialogue with Himself | 61
God Blessed Them | 61
The First Five Days of Creation | 62
The Creation of Humans | 63
Seventh Day and God's Rest | 64
All about Meaning | 65
The First Account | 65
The Second Account | 67
Conclusion | 73

∽ 3 ∾
The New Cosmic Story | 75

The Birth of the Universe | 76
The Travail of Galaxies | 77
The Significance of Stars | 80
The Birth of a Planet | 81
Light from the Past | 82

Cosmic Breakthrough—Living Forms | 83
Living Cells—A Wonderland | 89
Embedded Wisdom? | 91
Conclusion | 92

4
Understanding Our Place in the Universe | 93

Rediscovering a Cosmic Perspective | 94
Theory of Knowledge—A Very Brief Introduction | 95
Ancient Cosmologies | 97
The Cosmology of Classical Antiquity | 100
The Ptolemaic Cosmos | 103
Newton's Mechanistic Universe | 104
The New Cosmology—Toward Fundamental Wholeness | 107
Cosmic Fine-Tuning | 110
Cosmic Size Spectrum | 113
Human Roots and Culture in Cosmic History | 116
Conclusion | 120

5
Humanity—An Evolving Phenomenon | 122

Background | 122
Footprints in Tanzania | 124
Homo Erectus and Their Relatives | 126
The Earliest Cosmopolitan Homo and Their Successors | 128
The Rise of Homo Sapiens | 130
Developing Skills in Human Infancy | 133
The Dawning of an Interior Horizon | 135
Why We Cannot Ignore the Past | 138
Conclusion | 141

6
The Human Brain and the Rise of Human Consciousness | 142

Toward a Biology of the Human Spirit | 142
Consciousness of Place and Power | 146
The Human Brain—Complexity and Mystery | 148
The Brain and Its Neurons | 151
Brain and Belief | 154
The Brain's Influence on Religion | 157
The Brain's Influence on Conceptions of God | 158
The Spirituality of Children | 161
The Mystical Mind | 163
Consciousness as Inside Story of the Universe | 166
Conclusion | 170

7
The Inside Story Solidifies—Religious Symbols | 171

On the Way to Revelation: Symbol Recognition | 171
Perception of Symbols in Religious Experiences | 173
The Nature of Religious Experiences | 175
On the Structure of Religious Symbols | 178
Ancient Symbols of Divinity | 180
The Quest to Render Life Secure | 182
The Rise of Monotheism | 186
The Biblical Trajectory | 189
Jesus and the Rise of Abba Consciousness | 192
Christ: The Revelation of God as Love | 194
The Depth of the Inside Story | 196
Conclusion | 198

8
Desire, Sacrifice, and Revelation | 200

Desire as the Basis of Human Agency | 201
How We Became Human | 206
On Sacrifice, Scapegoats, and Victims | 208
The Mimetic Constitution of the Human Person—Homo Imitans | 212
The Crisis of God as Abba | 217
Ambrose (340–397) | 218
Augustine (354–430) | 219
Gregory the Great (c. 540–604) | 220
Anselm of Canterbury (1033–1109) | 221
Martin Luther (1483–1546) | 225
The Revelation of the Forgiving Victim | 226
Conclusion | 232

9
Toward an Expanded Theology | 234

Of Seashells and Pebbles | 234
Wholeness | 236
Subjectivity | 239
Inwardness | 243
Christian Inertia | 247
Abba Incarnate | 249
Trinity | 252
World without End?—A Concluding Reflection | 255

Bibliography | 261

Index | 271

Acknowledgements

THE ADVENTUROUS ATTEMPT OF opening a series of windows on the landscape of God's creation—so profoundly altered by recent scientific research—could not have been undertaken without the encouragement and help of many others. I gladly acknowledge their contribution and express my gratitude.

Among the first supporters of the project were several religious education teachers from a Christian high school who had engaged me after an address on a related topic. Although I have forgotten their names, I want to thank them for their candor when they freely admitted that their education had not equipped them to field frequent student questions on the difference between the scientific view of the creation and the traditional model. As my own understanding grew, first ideas for a book were shaped in conversation over coffee with two longstanding friends, Terry Craig and David Crockart. Thank you both for your generous engagement. Along the way, as Honorary Fellow of the Australian Catholic University, the project benefitted much from a generous access to literature and the opportunity to present papers at monthly faculty seminars under the leadership of Dr. Alan Cadwallader and his successor, Dr. Rapin Quinn. I cannot think of a better occasion than this to show my gratitude for collegial friendship and the freedom to explore ideas.

My acknowledgement of help received would be gravely deficient without paying tribute to Professor Emeritus Raymund Canning, who launched me on my career in academic theology, for his ongoing support over more than twenty years, to Professor Peter Gill for wise counsel, and to the fine work of the many silent scholars mentioned in the bibliography. You all have enriched my understanding beyond measure. I also want to thank those who engaged personally with parts of the manuscript: David Barling for his enthusiastic support and comments on chapter 4, Georgia Caminetti for commenting on chapter 3, Dr. Nicholas Coleman for reading chapter 6 and parts of chapter 9 and for comments and several helpful suggestions, and Professor Emeritus Ian Acworth for cheering me on at the right time. Many others—family and friends—stood with me during the often-solitary task. My special thanks go to Isabel Baldwin, David Cam, Christopher Clark, Ralph Hale, Marilyn and John McPherson, Mike and Carol Smith, Kevin Stork, Julia Tolmacheva, and Chloe Webster for ongoing interest and encouragement. I also wish to express my gratitude to the publisher, Wipf and Stock, for the smoothness of their processes and the cordial professionalism of the team. Thank you Jonathan Hill, Nathan Rhoads, and Matthew Wimer for the joy of working with you.

The largest debt of thanks I owe to my editor, Elizabeth Judd, and to my wife, Barbara, for giving more than I can say. Without them I could not have done it. Elizabeth has blessed the project with the outstanding talent of an insightful professional, refining the manuscript with precision and patience while remaining sensitive to the author's voice. Her meticulous work on the bibliography deserves special mention. What is more, when I lost confidence in what I was doing, Elizabeth knew how to walk me firmly through the valley. Lastly, gratitude of a deeper kind goes to Barbara, whose unflinching courage stood with me when a second cancer diagnosis threatened to cloud the horizon. The project was simply part and parcel of our perduring partnership. Thank you, Barbara.

List of Illustrations

Table 4.1. Cosmic Size Spectrum | 115
Table 4.2. Human Roots and Culture in Cosmic History | 117
Figure 5.1. Timeline of Human Evolution | 132

Introduction

An Autobiographical Note

ON PALM SUNDAY OF 1948, I knelt in the company of other 14- and 15-year-olds before the altar of St. Clemens Romanus, a twelfth-century Lutheran village church in Lower Saxony, Germany, to be confirmed in the Christian faith. In preparation, we had memorized the Apostles' Creed, passages from Luther's catechism, and the lyrics of some hymns. Somehow, I had believed in the existence of God since early childhood, but as I would discover years later, belief in God and living faith were not the same. After reciting the creed, we were to take Holy Communion for the first time. To remind us graphically of what was at stake, a richly carved triptych depicting Christ's crucifixion towered on the altar, the symbol of God's costly grace, while the rib-vaulted and frescoed ceiling of the choir detailed the final act in the drama of salvation. Christ, the judge of the world, was enthroned at the apex; to his right, music-making angels accompanied the elect to the City of God, while to his left devils with pitchforks drove the damned into hell's gaping maw. Lifting their gaze to the ceiling above the altar scene, beholders were confronted with the display of ultimate reality, emotionally dominated not by the joys of heaven but by the horrors of hell.

While such an analysis was beyond me at the time, I understood those threatening symbols viscerally all too well. Traumatic memories were still too fresh: the bombing raid that destroyed the family home and killed most of my classmates in 1943, our failed attempt to flee from the Red Army in 1944, the years of Soviet military rule that brought frequent nocturnal home invasions from intoxicated and looting soldiers, the forced deportation from the eastern German provinces, ceded to Poland by the Allies at the Yalta Conference of 1945. For years, shelter and safety always hung by the flimsiest of threads, ever vulnerable to being cut off without warning by some arbitrary power. Besides, in postwar Germany at that time the symbols of endured trauma were omnipresent—ruined cities, displaced people, cramped living quarters, makeshift classrooms, freezing temperatures in winter.

On that memorable Palm Sunday, I had little understanding of how deeply these experiences had colored my outlook on life and on God. At a gut level there was fear and anxiety, even as I waited to partake of God's covenantal meal for the first time, and the frescoes above were of little help emotionally. Little did I know then that these works of religious art were themselves products of a profound religious anxiety that pervaded the Christian world throughout the Middle Ages, during the Reformation, and for centuries beyond. Certainly, the solemnity of the present moment was not lost on me, although its meaning as a living reality was still a long way off. Many more years needed to pass before *this* meal met a deeper hunger, when it mediated to me its full significance not only as a symbol of belonging and peace but as an abiding existential reality.

Yet I still remember most vividly how, back then, I had turned earnestly toward God, perhaps expecting that such an inner exertion would make God somehow more real to me and me more acceptable to him. Perhaps I'd hoped that God would reciprocate and confirm through some discernible experience or personal revelation that I was wanted. But was such an expectation not illusory from the judge of the world? Whatever the motives, it was the conclusion that my mind began to form, tainted as it was by the ample experience of rejection that set the course for the next 30 years. Since nothing identifiably divine happened on this day or in the days that followed, I took it to mean that I was probably not among the elect. If so, I was on my own with my fragile hopes, fears, and failures, yet determined to put on a brave face, hiding my fragility and struggles with life's groaning and growing perplexities under a mask of self-sufficiency.

In the 1950s, my family relocated to North Rhine–Westphalia, a heavily industrialized and coal-mining region of West Germany, which offered new opportunities in the upswing of the country's reconstruction. After graduating from high school, I studied geoscience and cartography, initially not quite knowing what attracted me to this choice. Only in later years, as I got to know myself better, did I grasp the underlying pattern.

What motivated me were occasions that allowed me to explore, comprehend, and bring to light what was hidden and mysterious, and then by connecting the dots create physical or conceptual maps that depicted and explained what I had found. Thus, a perpetual search for coherence turned out to be my calling. Professionally, this work would take me, together with my late wife, to Egypt, India, West and South Africa, Saudi Arabia, Colombia, the United States, and eventually to Australia. There, in 1977, almost 30 years after my first overt longing for God, God's love apprehended me in a profound and personal experience.

On an exceptionally hot summer's day, I was mowing the grass around our house in Sydney when, amid the din of the mower engine and slashing rotary blades, a thornbush to my left riveted my attention. It had been there all along, but this morning it strangely mattered. In a vivid flashback, I saw myself, 11 years old, in the abandoned farmhouse where our family had found temporary shelter from the Red Army. In the trash that greeted us inside, I spotted a Bible. As I opened it, Christ's bleeding head confronted me from the book's etched frontispiece. As this thorn-crowned memory broke into my consciousness as clear as it was so many years ago, an inner voice—unperturbed by dust, noise, and perspiration—addressed me: "Come and follow me."

By that time, I had lived for decades on the far side of Christianity. Yet, its symbols had not lost their meaning, and I knew instantly who was speaking. Professionally, I was at the height of my career and a course change was not on my radar. For several years I had been deeply involved in the Australian uranium industry, having negotiated one of Australia's first commercial uranium-supply contract—a contract worth several hundred million dollars—with two Japanese nuclear power companies. Now, in the midst of my pursuits, the voice stripped me of my professional certitude, opening an inner wound that could not be staunched.

Surprised by a sudden hunger for spiritual knowledge and understanding, I began to read earnestly and passionately. Thomas Merton's autobiography, *The Seven Storey Mountain*, as well as his other works spoke deeply to my awakening spirit, together with the works of other spiritual writers, especially their reflections on the Scriptures. Eventually, I knocked on the door of the local Anglican church, where I falteringly began my Christian walk in a tradition that provided the communal context and spiritual discipline I craved. Ravenous for reality and truth, I wanted what the gospel had to offer, no matter the cost. In gratitude for grace so freely given, I had lost all appetite for critical inquiry. Spiritually,

I had much catching up to do. Only years later, in a conversation with a group of high school students wrestling with how to relate the teachings of their biology class to their faith, was I forced to think afresh about the doctrine of creation and its relation to modern science, and a new journey began.

This journey continues to this day. After my retirement, I studied theology formally at the Australian Catholic University in Canberra (MA theology, 2002; PhD, 2006). At this "boutique campus," I found much freedom as a Protestant to explore critical questions and space for reflection. Those studies were followed by a part-time lectureship for two years and an eight-year honorary research fellowship. That cherished eight-year opportunity for research, stimulating discussions, and collegial friendships led to the publication of several book chapters and peer-reviewed articles in theological journals. It also provided much of the inspiration for the writing of *Cosmos and Revelation*.

How the Book Came to Be Written

Having explored the natural world as a professional without knowing its Creator, I returned to this captivating subject with an entirely new perspective. This frame of mind was grounded not only in the love of knowledge as such but in love for the one whose ecstatic charity I believe creates and sustains all there is.

However, there was a catch. On the one hand, Western civilization had relied for centuries on the traditional reading of the biblical creation narratives (Gen 1–2) as a blow-by-blow historical account of how creation had occurred. On the other, because of discoveries in cosmology, developmental biology, and other fields, the current scientific understanding of the natural world has diverged profoundly from those creation narratives. Yet millions of Christians still cling to the biblical account. Since this difference over the doctrine of creation is deep and is prosecuted with hostility wherever it surfaces, Christianity can no longer offer a coherent narrative of the cosmos.

The players involved in this troublesome conflict can be grouped by their degree of alignment with the scientific model:

- At the conservative end of the spectrum, we find the *young-Earth creationists* (YECs). These Christians insist that the text of Genesis 1–2 is a reliable historical account of how creation occurred step by step less than 10,000 years ago. Many YECs are aware that these beliefs contradict scientific findings, yet they reject these findings on the basis that biblical authority trumps scientific discoveries. Some even think that scientists deliberately manipulate or fabricate the data to undermine belief in a Creator, while others believe that the data of science can be used to support their claims.

- Another group of believers, or *old-Earth creationists* (OECs), are prepared to accept the scientific evidence for Earth's old age. But, rejecting Darwinian evolution, they insist that the diversity of life on the planet was the result of direct divine action. In other words, creation may have taken place over long periods, but God brought forth the diversity of life we see today by direct intervention. Like the first group, OECs insist on the historicity of Adam and Eve.

- The third faction in this conflict over how God created life on the planet are adherents of the *intelligent design movement* (IDM). They believe that all created reality is characterized by an "irreducible complexity" that calls for the presence of an "intelligent designer." Like the first two groups, IDM proponents deny Darwinian evolution. By calling the designer God, their theology is a "natural theology" that attributes the observable complexity in the natural world to special divine interventions.

- The fourth group aligns itself fully with the findings of modern science. They are called *theistic evolutionists* (TEs) and believe that God creates by means of processes observable in cosmic, biological, and cultural evolution. Most TEs understand Adam and Eve not as historical individuals but as symbols of humanity. While some believe that God providentially guides these processes to bring about his purposes, others, like myself, have adopted a more expansive and theologically more coherent interpretation: the Creator has brought into being a creation with the built-in freedom and creativity to create itself. Before elaborating on this view, on which *Cosmos and Revelation* is premised, let me cite a fifth group that, although rarely mentioned, has a significant stake in this conflict.

- The fifth group encompasses those that could be called *believers in transition*. They have become increasingly skeptical of YEC claims that the scientific model of the cosmos and its evolution is false; they recognize that Western culture, now gone global, relies on science and technology to fly from New York to London, develop sustainable energy sources, implement Mars landings, and create vaccines against contagious diseases. They have become convinced of the need for an updated understanding of God's creation; they seek to become more knowledgeable about scientific discoveries and, given the grace of God, to test what they may be prepared to accept as true. Despite their courage to break new ground, these believers, especially students and young scientists, need pastoral support and encouragement, reassuring them that in the quest for intellectual integrity in the age of science they are not alone—an important motive for the writing of this book. Hungry for wholeness, these believers in transition will find little help in churches that lean toward the conservative end of the debate, which often includes the outright rejection of evolution as an instrument in God's creation. Pastorally speaking, they are unlikely to feel supported unless the Christ their churches proclaim is large enough to embrace and redeem the vast and still-expanding cosmos the sciences have discovered.

With these thoughts in mind, let me return to my own position in the creation-evolution debate with a six-point summary:

1. When primordial energy vibrations first flared forth, God's creative love overflowed in power, creating all that exists potentially, including a capacity to improvise at the highest level of ingenuity, allowing the creation to explore all realizable potentialities.

2. This act involved God in the risk of contrariness. Had God wanted to avoid this risk, he could have created a static world in which everything remained the same. Only a creation free to be itself makes sense theologically, for only such a creation can respond to the God who created it in love. I accept the idea of a "fallen world" in the sense that the creation had to "fall out of equilibrium" after the big bang because an expanding and dynamic universe such as ours could never have come forth from a stable state which is the enemy of change.[1]

---

[1]. This touches on the question whether the storyline of a human fall with creation-wide consequences—as traditionally understood—needs to shift with the discovery of an ongoing and fundamentally dynamic creation. Thoughts about the direction of this shift may become clearer in chapters 5 and 8, where I consider the late cosmic appearance of the human as an evolving phenomenon. To offer a hint, the unfolding of the creation and the perception of God's presence within it was never a straight line.

3. When God created this other, this act included the genius by which the universe is able to explore all the potential forms of organization contained in it by way of observable randomness at the microlevel, as in biological evolution. These appeared in space-time from the God-endowed interplay of chance and law, explaining why life on Earth has progressed through a complex history of innovation and change.

4. With the rise of human consciousness an enormously significant element has appeared. From now on, personal subjectivity resides in the universe as the *inside story*. While science cannot measure its throbbing pulse in cosmic history, this inside story is told nevertheless based on the experiences of personal subjects. Unthinkable without the *outside story* of phenomena that evolved over billions of years, this insideness is potentially the seat of divine revelation and personal devotion.

5. Since we are still living in this moment of creation and are part of its birth pangs as the creation is moving towards its completion, we, as God's cocreators, are aiding or hindering this process to which every person contributes.

6. Understood from the horizon of the resurrection, we are shown not only the inconceivable strength of divine love for this one human being, revealing God's loving closeness, but also the final uncoupling of God from any human representational capacity. In this light, the future comes to us not as a bland "everything is possible" or as a vacuous "where to next?," but as a gift. It summons us humans to faith-based creative action within the realistic possibilities of our given historical situation, which includes the emergence of a culture informed by a scientific description of the world.

This position conceives of God's "radical immanence" in the evolving creation as creation's inside story that hides God's transcendence, even the Lamb slain from the foundations of the cosmos (Rev 13:8).

The natural order as the sciences describe it is bound to prompt several questions for Christian faith. For instance: about God's relationship to the scientific age in which we live; about Christian responsibility

---

Instead, it was a groping and gradual awakening to rightness against the potential for contrariness inherent in the initial condition of the universe. The presence of contrariness is thus not the result of an unfortunate misstep and a divine curse, but the occasion for the revelation of divine love.

to culture; about the credibility of Christians when they take advantage of modern science and technology (medicine, aviation, communication, etc.) but reject the findings of cosmology and developmental biology; about the process-nature of creation, especially advances toward novelty and the way Christians do theology; about the misrepresentation of God as creator when Christians deny scientific findings (e.g., evolution and big bang cosmology); about the need for a creative expansion and reexpression of some key doctrines (e.g., creation, original sin, incarnation, atonement, resurrection, new creation) under the auspices of an evolutionary natural order; and about a deeper understanding of the way the body of Jesus of Nazareth was made like ours, atom by atom, of the same stuff of the universe, thus embodying all of cosmic history.

In the age of science, questions like these are not only unavoidable but also of great significance, as they point to the theological task before us if we want to meet the spiritual hunger of our time. I caught a glimpse of this significance about a decade ago, when I initiated a citywide cross-denominational survey of churches in Canberra, the national capital of Australia. A colleague from the ACU Faculty of Education, Sue Wilson, and I wanted to discover the degree of openness and resistance to the new world picture the sciences are presenting. The defining issue turned on the presence of two distinct subpopulations in the sample: respondents considered Open (64 percent) and Resistant (36 percent).[2] Nearly half of the Os viewed the sciences positively because they rely on "verifiable evidence" (48 percent) and "glorify God" (43 percent), while more than half of the Rs resisted because scientific findings disagreed with the Bible (55 percent) and theories of science could not be trusted (47 percent). Despite these conflicting measures, nearly 70 percent of all respondents felt that the churches needed to come to terms with scientific findings for the sake of their mission in the twenty-first century.

Given this tension, church leaders tend to avoid the subject, to which my own experience of church life testifies. Over the last 35 years, I cannot recall one single instance when the Christian doctrine of creation featured as a teaching or preaching topic. As I cannot imagine that church leaders, ordained and lay alike, are unaware of how deeply the discoveries in cosmology, physics, biology, genetics, and paleontology have shattered all traditional conceptions of space and time as well as of animal and human history, there is only one explanation: they don't

---

2. Anglicans constituted 50 percent of Os and 26 percent of Rs.

know how to deal with it without causing intracongregational conflict. The fact that clergy education also ignores the new-universe story does not help. Hence my contention that faith in God as creator of the natural world lacks credibility without due acknowledgment of its contemporary scientific description.

The earliest impetus for writing *Cosmos and Revelation* goes back to several unpublished presentations given in research seminars at ACU's Canberra campus between 2010 and 2015 under the inspirational guidance of Dr. Alan Cadwallader and to the lively collegial discussions that followed. As noted earlier, that warm fellowship and intellectual stimulation helped inspire the writing of the book.

Now, I offer it with the hope that the Christian community may lay down its unwarranted opposition to scientific findings; that it may begin to see the challenges posed by science as the theological frontier of our time, beckoning us to move forward with courage, assured that God is deeper than Darwin; and that, if trinitarian love has anything to do with the new universe story, we will find the wisdom to guide our creative reflections and spur us to move deeper into the unfathomable mystery we call God.

On a personal note, since I am writing more as a teacher and seeker of wisdom than as a systematic theologian, the book does not set out to prove a thesis. The windows on God's creation I am opening are meant to help readers to intuit for themselves something of the immensities God's self-revelation in Christ must encompass for the doctrines of creation and recreation to make sense in an age of science. As to method, I have been sharing more a process of discovery than solutions.

Obstacles Ahead

For many people, reading this book will mean entering unfamiliar territory: habitual themes will be transposed into a new key as old paradigms give way to a new description of God's creation under the weight of scientific evidence. A creation that once was thought to be made of a collection of fixed objects standing in external causal relations to each other turns out to be constituted by intricately dynamic and interrelated processes exhibiting both lawfulness and randomness. A creation that once was understood as a finished work turns out to be ongoing. A human species that thought of itself as a separate entity turns out to be an integral part of the natural world. In short, the new, grandiose story of the creation presents the Christian religious imagination with a radically altered landscape. We need to understand that although this new picture of the world challenges some cherished facts considered unalterable by the Christian tradition, we are seeing for the first time how the creation was constituted—a fact that has been hidden for millions of years whose discovery does not challenge the glory of God. Indeed, this moment of unveiling or revelation in world history is not unlike that which happened to the once-reigning flat-Earth paradigm when in late antiquity Greek astronomy discovered a spherical Earth for the first time.

With time, a huge gap between the transcendent dimension of human existence and the physical reality as science describes it has developed, not least because Christian theology since the Middle Ages has not kept pace with science. Consequently, many thoughtful believers find themselves overtaxed when called on to relate Christian claims to the new paradigm of creation, and most clergy feel unqualified to discuss the subject. From early on, Christianity has thought of the natural world (usually limited to planet Earth) as the stage on which the drama between God and humanity is played out. Since that stage was fixed, it was taken as a mere backdrop for the drama, warranting little further attention. This scheme placed the primary focus on the personal relationship between God and humanity, leaving the cosmos out. This conception of things fitted neatly with a Christian anthropology that considered humans a special entity existing not only separate from the natural order but in a state of alienation from it.

As it is, the scientific account presents us with evidence that death is a natural phenomenon as old as the first living cells that appeared on the surface of the Earth some 3.6 billion years ago. Their fossils were detected in rock formations of the Pilbara Craton in the North of Western Australia, one of only two pieces of pristine Earth's crust from the Archean Eon, 3.6—2.7 billion years old, identified on the planet. We also know that predation, diseases, lethal genetic mutations, extinctions, and natural disasters like floods, earthquakes, and ice ages occurred long before humans (*Homo sapiens*) came to live on Earth. In a world that increasingly prioritizes evidence over revelation, such discoveries precipitate an unbridgeable credibility gap for the Christian claim that *Homo sapiens* was the source of the world's disorder, in itself a sign of God's judgment on account of Adam and Eve's disobedience in the garden of Eden.

Without brushing aside the limitations of science, its methods have been singularly successful in revealing what is legitimate for science to discover, and there seems to be no stopping it. For instance, it has shown us that in the quantum universe we inhabit everything is related to everything else, that the entire cosmos is in process, that no system possesses properties except through interaction with others, and that in the new story nothing looks the same as before. In other words, when we read the new story, we must be prepared for change!

It goes without saying that many theologians and believing scientists have been aware of these challenges, though mainstream church thinking has yet to acknowledge them as such. As most Christians have been taught as one of their fundamental assumptions that the creation was a fait accompli, they find it hard to embrace the creative-process nature of all that exists. Yet, the weight of this new dynamic context keeps nudging the Christian imagination to expand its horizon toward a fresh, exhilarating vision of God commensurate with the opulent creativity and grandeur that the new cosmic story so abundantly displays. Far from pushing God out of the universe, the new cosmic story adorns and magnifies the Creator of this vast and breathtakingly wonderful creation.

The Challenge for the Churches

Assuming the churches incorporated the scientific view of nature into their theology, numerous, even daunting pastoral and didactic challenges are bound to arise. This should not be surprising. On the one hand, despite decades of exposure to the new cosmic story, church leaders and seminaries have tenaciously clung to prescientific premises, as if Charles Darwin had not been born and Galileo had not invented the telescope. On the other, questions arising from the new story are inescapable. For instance, if God's creation is indeed as the sciences describe it, wouldn't this alone demand that the Christian community take the scientific description seriously? And further, wouldn't the failure to do so simply reinforce the trend of losing voice in public affairs? How much credibility would Christian claims gain if church members were able to speak competently about the compatibility of the new cosmic story with the Genesis account, or if they were able to explain the cosmic and spiritual significance of the late arrival in cosmic history of creatures like us? As it is, the voice of the church is today culturally so feeble that it is barely heard.³ In an essay on "Science and Religion in the United Kingdom," Christopher Southgate observed that "if the Church wishes to have a voice in shaping society, it will . . . have to earn that voice with demonstrable expertise."⁴

My impression is that mainstream Christian theology seems to have no appetite for tackling the problem, preferring to continue as it has for the last 500 years. Both Roman Catholic and Protestant theology have focused exclusively on God and the individual since the Reformation as if the rest of creation did not matter. This neglect, Catholic theologian Elizabeth Johnson pointed out, has enfeebled the discipline and prevented the emergence of a broad-based Christian commitment to ecological concerns.⁵ For instance, when it was reported some years ago that during a single human lifetime the global wildlife population had declined by 60 percent, the church did not respond, while a well-known climate-change- and science-denying evangelical church leader proclaimed in a sermon full of scientific misinformation that God intended the Earth to be a "disposable planet."⁶

Among Protestant leaders, one of the early pioneers who addressed such concerns was the renowned biochemist and theologian Arthur Peacocke (1924–2006). As an Anglican priest and chaplain of Oxford University, he contended as early as 60 years ago that, by hiding behind dogmatic pronouncements, the churches were trying to prevent science from challenging Christian truth claims. Convinced that such a stance could not succeed, Peacocke devoted much energy to bringing change. He argued that the dissonance between church language and the way people perceived the world was disaffecting more and more people, while an underequipped leadership (ordained and lay) was unable to meet the adaptive cultural challenge posed by an increasingly scientifically oriented public. He warned that unless this trend was corrected, the

---

3. The loss of this competence may be illustrated with a news item (April 23, 2021) reported by one of Australia's Christian media outlets. In ethical decisions about advances in bio- and health technology, Australian religious leaders scored second lowest in public trust (12 percent, just ahead of politicians, 6 percent). By contrast, doctors and scientists ranked highest with 77 and 73 percent, respectively. https//:www.eternitynews.com/who-do-Aussies-trust-to-sort-out-ethical-issues-scientists-and-doctors/.

4. Southgate, "Science and Religion."

5. Johnson, "Presidential Address."

6. Professor Emeritus in chemistry, University of Glasgow, Paul Braterman, "God Intended," http://theconversation.com/god-intended-it-as a-disposable-planet-meet-the-US-pastor-preaching-climate-change-denial-147712.

Christian community was in danger of becoming an esoteric society that communicated only with itself.

Six decades later, the state of Christianity in Western Europe looks astonishingly like Peacocke's prediction. The heartland of the Protestant Reformation and the place where Catholicism had its home for most of its history is today Europe's most secularized region.[7] My own experience suggests that Western culture is no longer at home in the cosmos as the sciences describe it—churches, their leaders, and theologians included.

Most live without much thought on the leftovers of an outdated cosmology that provides neither a sense of cohesion nor of belonging. In liturgy and creeds, the old cosmology remains deeply embedded in Christian consciousness. If Christians think about the universe at all, for most of them it is little more than a vast container filled with stellar objects to which God is external. When theologians speak of "the world," they invariably mean not the universe but planet Earth, unconcerned that they unwittingly communicate a cosmological conception that is several hundred years out of date. When it comes to evolution, churches tend to avoid the subject as too controversial, simultaneously implying that it is irrelevant or marginal to our lives. Yet, scientifically speaking, evolution is simply the way biological, chemical, physical, and cultural systems function as they adapt to changing conditions with creative novelty. Seen with the eyes of faith, these capacities take on a God-endowed character.

What is more, the churches have yet to realize that they live in an unfinished world, a quantum universe going through the juvenile stages of cosmic emergence in which neither space nor time is an absolute, and where nothing is fully determined. Yet, many Christians have written the cosmos off because "Jesus is coming back soon" (only to Earth?), not realizing that such an attitude deprives the creation of its ongoing partnership in the gospel and of caring attention. Moreover, this stance sanctions the absence of a unifying cosmological-theological vision, which in itself is a major source of our global problems.

Another stance that Christians have adopted in matters of science may be illustrated by a (hopefully fictitious?) story about Michael Faraday, the famous nineteenth-century physicist. It goes like this: "When

---

7. Pew Research Center, *Being Christian*, https://www.pewforum.org/2018/05/29/being-christian-in-western-europe/. While some, but not all, of these losses are attributable to the death of cultural Christianity, the observations of this survey should give us pause.

[Faraday] went into his laboratory, he forgot his religion and when he came out, he forgot his science."[8]

The problem with this dichotomy is that it can only be sustained at the price of intellectual integrity. It not only contradicts a core tenet of our faith, namely that God created *one world*, but it also blinds us to what is so remarkable about the creation: the emergence of cosmic insideness through the rise of human consciousness, the vessel of divine revelation in the world. In terms of this book, "cosmos" and "revelation" cannot be separated, as affirmed by Paul long ago: "For since the creation of the world God's invisible qualities—his eternal power and divine nature—have been clearly seen, being understood from what has been made" (Rom 1:20).

A significant motivation for writing this book was the desire to help contemporary Christians grasp afresh this one-world perspective. It is my hope that by immersing themselves in this awesome wholeness, readers may regain some of the intellectual integrity forfeited in the seventeenth century when Thomas Hobbes declared that the world was nothing but a thoroughfare of atoms in motion and when his contemporary, René Descartes, proclaimed the duality of mind and matter. Sadly, in their unresponsiveness to this nearly 400-year-old problem, the churches are reinforcing this false dichotomy, which, in my view, can only lead to a growing loss of cultural credibility by Christianity in our time. Regaining a cultural voice requires at a minimum that Christians forsake their combative rejection of evolutionary science and become constructively engaged with the new cosmic story. *Cosmos and Revelation* was written with this need in mind, hoping to present this story in a way that stretches the mind and opens the heart. The sweeping arc of its plot across key scientific discoveries speaks implicitly of wonder and beauty as well as of pioneering courage and hope, inviting readers to step outside traditional conceptions and trust a faithful Creator who is behind all our seeking and longing for coherence.

Affirmations, Limits, Advocacy

This introduction would be deficient if it did not say a few words at least in broad outline about what the book affirms and what it does not intend to explore.

---

8. Polkinghorne, *One World*, 97.

The fundamental premise of the book is that the triune personal God of the Christian confession is the creator of all things "visible and invisible" in the micro- and macrorealms of creation's immensities as discovered by modern science. Their vastness immeasurably exceeds all previous conceptions and traditional descriptions of the creation, especially those derived from a literal understanding of the Genesis accounts of origins.

To be sure, the book's cosmic scope will stretch the theological imagination, sometimes painfully so because the churches have sanctified the medieval model of creation, limited as it is to no more than the solar system: Earth beneath our feet, celestial bodies above, and beyond their spheres the throne of God with Jesus Christ, the judge of the world, at his right hand. In the new cosmic framework of 100 to 200 billion galaxies the size of the Milky Way, each with at least a 100 billion stars like our sun, and with a far future measured by tens of billions of years, this geocentric and anthropocentric conception of the universe becomes unsustainable. Even Teilhard de Chardin's brave pioneering vision remained irreversibly beholden to his theological education that the fate of the cosmos turns on what happens on planet Earth. Put simply, Christian theology must step away from this fallacy because theological ideas are not without consequences for the world. How else will we sustain a vision of the cosmic Christ within a cosmos that forces us to take seriously the possibility of extraterrestrial life in countless places?

While I have taken the principle of evolution deeply into my conception of the creation, in the sense that biological evolution is the organic subset of cosmic evolution, I do not offer a description of evolution as such in its functional detail. Although I speak of cosmic "inwardness" and "subjectivity" as created capacities capable of linking divine influence and creaturely response, revelation included, I have otherwise not attempted to peer behind the nature of divine action in creation. The themes of inwardness and subjectivity have been developed further in the trinitarian section of chapter 9. In that section, I emphasize God's action in creation out of freedom and love, God's continuing personal and substantial presence everywhere in, yet without imposing his purpose on, his (ongoing) creation.[9]

Lastly, the book does not accept the logic advanced by many antievolutionists arguing that a world with defects contradicts the perfection

---

9. Although subjectivity and creaturely response are intimately related to questions about miracles and theodicy, I have not attempted to address them.

of creation, forcing a redefinition of God's goodness, and that an evolutionary approach eliminates Adam and reduces the parallel with Christ (Rom 5) to a metaphor.[10]

Conclusion

The book's primary purpose is to invite gospel-believing Christians to embark on a journey of discovery by looking at a set of scientific discoveries with open eyes and without prejudice. Each one of these calls for serious theological reflection if Christians want to interact coherently with the age of science. I offer this work as a modest contribution to the advancement of a Christian vision for our time, aware that if such a vision is to emerge, it will need an atmosphere of generosity, breadth of scale, and inner coherence. Pierre Teilhard de Chardin, whose work I admire, was convinced that "the bigger the world becomes and the more organic its internal connections, the more will the perspective of the Incarnation triumph."[11] In the same visionary mode, Teilhard anticipated that believers, although at first afraid of evolution, will eventually discover that evolution offers Christians a magnificent means by which to grow in their union with God, whose love drives this process toward ultimate convergence. For now, you the reader are invited to share his perception that "science alone cannot discover Christ . . . [but] Christ satisfies the yearnings that are born in our hearts in the school of science,"[12] for cosmos and Christ are inseparable.

---

10. Since revered colleagues in the field of theology and science have done likewise, I find myself in good company and refer the reader to publications by Ted Peters, Robert John Russell, Christopher Southgate, Bethany Sollereder, and others.

11. Teilhard de Chardin, *Phenomenon of Man*, 325.

12. Cowell, *Teilhard Lexicon*, 172.

1

The Nature of Revelation

MODERN SCIENCE HAS CREATED formidable challenges to traditional views of the world, how it works, and how it came to be. These discoveries raise completely new questions for theology and the concept of revelation, challenging Christians to address them if the message of the church is to meet the spiritual hunger of an increasingly science-informed public. By exploring various approaches to the meaning of revelation, this chapter sets the stage for what follows.

The Dimension of Mystery

The oldest fossil of the genus *Homo*—known as the "Jedi jaw" and found in Ethiopia—dates from around 2.8 million years ago.[1] The last shoot on the human family tree had begun to stretch forth at least 100,000 years before the present as a new species, *Homo sapiens* (the self-styled "wise human"). With rising consciousness, their epic voyage of self-discovery was accompanied by a growing awareness of a mysterious dimension that encompassed their existence. It met them again and again in their living and dying, in the natural world with its wonders and dangers, in pain and suffering, in the forces that formed the land, that brought or withheld rain, in the awesome vistas of the night sky, and more. Already their older

---

1. Smithsonian National Museum of Natural History, "Oldest Fossil," https://humanorigins.si.edu/research/whats-hot-human-origins/oldest-fossil-our-genus.

cousins, the Neanderthals, with whom *Homo sapiens* lived and interbred in Europe between 40,000 and 30,000 years ago, seemed to have had a remarkable awareness of mystery. Based on a Neanderthal find dated from 176,000 years ago, the Bruniquel Cave in the Aveyron Valley (France) holds a rock structure erected from 400 deliberately broken pieces of stalagmites in two concentric rings. Dubbed the Paleolithic Stonehenge, it seems to have served cultic purposes.[2] Sacramental behavior has been observed elsewhere, in ritual burial sites dating from around 100,000 years ago, and in an even older *Homo erectus* site. What these finds seem to indicate is that the human response to the mysterious dimension is religion.

As speech developed along with the ability to symbolize, our prehistoric ancestors told mythical stories about their world. In Australia many Aboriginal groups, though widely scattered across the continent, share variations of the same myth that tells the story of a powerful, often dangerous serpent of vast dimensions. This mythical creature is believed to be a descendant of the much larger being visible at night as a dark streak in the Milky Way. It is closely related to rainbows, rivers, and deep water holes. The Rainbow Serpent, according to legend, reveals itself to people on Earth as a rainbow, as water, as floods; it names springs and water holes, swallows people, and sometimes drowns them. It also imparts knowledge and wisdom, enabling some humans to make rain and heal illnesses, striking others with weakness, sores, illness, and death.

The practice of naming the mystery and telling stories about it belongs to the realm of primal religion. Our ancestors lived in an enchanted world and by naming the mystery its invisible forces not only acquired less fearsome and fateful forms but became more accessible. With time, belief systems emerged, including moral codes, rituals, and norms for communal living. In early religion, symbols *are* the actual presence of the numinous mystery, not a mere representation, almost a type of revelation. When Aboriginal people, for instance, perform their ancient rituals, they are living the past, not symbolically but actually.

As we would expect, religions that have come to prominence in the historical period—the period dating from the invention of writing in the fourth millennium BCE—are known as historical religions. Here, participants are more conscious that the mystery they are aware of is not the same thing as the symbol. A statue or picture of a god no longer stands

---

2. Yong, "Shocking Find," https://www.theatlantic.com/science/archive/2016/05/the-astonishing-age-of-a-neanderthal-cave-construction-site/484070/.

as its numinous power but as its symbolic representation. With advancing historical consciousness, people became more conscious of one all-encompassing, infinite, and eternal power in the world that, in the end, is perceived as incomprehensible. They know that no statue, image, ritual, or moral code can adequately capture the mystery, for what is infinite and eternal must always be beyond what is merely finite, representational, and symbolic. If this mystery is to be known, it will have to disclose itself.

However, symbols bring a certain problem. As descriptions, depictions, and representations with all their sensual appeal, they can be misleading and even dangerous. Having emerged from within a given culture, symbols tend to make people comfortable with their religious tradition, even indulgent in that the need to wrestle with ultimate questions in each historical situation is not urgent. Speaking in the current Christian context, theology has been too comfortable with its traditional answers, especially regarding creation. Today, scientific descriptions of the universe have so radically altered the cosmic and societal perceptions of the world that a new Christian understanding of creation is required. Christians and Christian theology must come to terms with the expanse of creation, its complexity, its incompleteness, its emergence over time, its inner dynamism and wholeness, and with humanity's place in the cosmic and biological order. Simply put, to the degree that the Christian vision of creation remains tied to the old cosmic story, it is too small. The same goes for a vision of the Creator.

Still, many Christians reject the findings of modern science as a basis for their understanding of creation; having chosen the so-called plain reading of the Genesis text as the only valid exegesis, they are captive to an outdated cosmic conception that inhibits their ability to communicate meaningfully with contemporary culture. By the same token, many who have embraced the scientific description of the cosmos struggle (as I have) with their commitment to remain faithful to the scriptural text. It is mainly for them that I have written this book.

The book is based on my belief that this world, as the sciences describe it, was created in God's perfect love as revealed in Jesus Christ, that it reflects the best possible world, and that it is sustained in being by God's will. It is further based on the understanding that creation includes multiple levels of biological organization, and that one of these brought forth a level of human consciousness that resulted in the present age of scientific discovery. Additionally, since this love-created world is far larger and more dynamic than anything Christians have ever imagined—the

Milky Way, our home galaxy, alone registers as only one among hundreds of billions of other galaxies, each containing 200 billion stars like our sun—it seems to me that we urgently need a vastly expanded vision of the Creator. How else would Christianity in an age of science make plausible its claim that God revealed himself in human form—the result of a long and arduous cosmic process over billions of years—as God's ultimate way of expressing his character as creative love and as the ultimate mystery?[3]

A Brief Theology of Revelation

In the broadest sense, theology is critical thinking about religious beliefs and more specifically also the science of the God of revelation. General parlance often uses "revelation" when something that has been unclear or obscure to us suddenly makes sense, or when, in a moment of illumination, we can somehow see reality more clearly than by ordinary reason alone. Yet many puzzles remain. For one, revelation as generally understood is personal. Theologically, it is an occasional disclosure of divine truth through the experience of a single individual like a prophet or seer, but this disclosure is not obvious to everyone.

Then there are questions about its origin. Culturally, there is a deep-seated notion that religious experiences that mediate ideas, say about the meaning and purpose of the world or human lives, are derived from "above," from otherworldly sources—that is, by way of revelation—while critical reason asks, "How do we know?" A secular world may say that moments of illumination are nothing more than the effect of our brain cells connecting the dots in a new way. And even if we hold to a divine origin of revelation, we cannot avoid the problem that such knowledge is thoroughly conditioned by nature, as it needs to make use of our neuronal tissue, while many are convinced that such a view would demean the value of revelation. Yet, we must go farther and ask how the concept of divine revelation is to be understood in our time.

Then there are questions regarding God's revelation in history. How plausible is the claim of the Judeo-Christian tradition today that God should have been so partial as to reveal himself to one particular man, Abraham, at a particular time in history and later to his offspring, the

---

3. While one can affirm a general correlation between Protestant values and the values of science, conservative Protestants have by and large been resistant to the discoveries of science (e.g., theories of relativity and evolution), whereas members of the Catholic tradition have more readily, albeit critically, embraced them.

people of Israel? What these questions highlight is that the notion of revelation, while on the one hand a rather simple notion, is on the other hand not easily dealt with in our time.

The word *revelation* comes from the Latin verb form *revelo*, meaning to unveil (equivalents in Hebrew and Greek are *gālâ* and *apocalyptō* respectively), all expressing the same idea of unveiling something hidden. The Bible uses this vocabulary in several ways: making obscure things clear, bringing hidden things to light, showing signs, speaking words, and causing the persons addressed to see, hear, understand, and know. In the Hebrew Bible the words used have no special theological reference, while in the New Testament *apocalyptō* and *apocalypsis* are exclusively theological.[4]

The Bible assumes that for humans to know the transcendent God, God must take the initiative and disclose himself. Moreover, humans must depend on God's revelatory initiative because humans are sinful, implying that the human mind is so prepossessed by its own ideas that humanity's natural powers are inadequate to understand or search out God.[5] Because revelation is God's personal self-disclosure, the encounter with Mystery is as with a person who offers fellowship and lordship. Since the encounter is with a person and not with an object, what is offered is to be thankfully received and responded to in an act of trust. These aspects take the experience beyond what we call the subjective in the ordinary sense, for what this encounter evokes lies outside such classifications. In fact, it is their radical antithesis. Let me explain. At the heart of biblical revelation is personal correspondence between what is revealed and faith. Since faith comes by reflection on how we personally respond to such an encounter, our response and correspondence also determine the notion of biblical truth, which differentiates it from any other form of truth.[6]

Revelation always has two focal points: God's person and God's purpose. God addresses people to tell them about himself, who he is, what he has done, what he will do, and what he requires them to do (see Adam, Noah, Abraham, Moses, Israel, the prophets, Paul, the apostles). At the same time, when God addresses people, he also confronts them with himself, but as a persona, who comes of himself to individuals to make himself known to them. If this sounds complicated, we must remember

---

4. Packer, "Revelation."
5. Packer, "Revelation."
6. Brunner, *Truth as Encounter*, 109.

that in its essence the concept of revelation is exceedingly simple; in Karl Rahner's terms, "Revelation means fundamentally the communication of the mystery of God to the world. The divine self-communication influences the world at every phase of its coming-to-be, and not just at the human level of propositional understanding nor within the confines of the biblical world alone."[7] Saying the same more concretely, John Haught writes: "Revelation is a constant, ongoing outpouring of God's creative, formative love into the world. In this sense it has a 'general' character, and in some way its every being is affected (and even constituted) by this universal divine self-communication."[8] Here we see how contemporary theology merges the idea of revelation with the biblical theme of creation. In other words, the self-revelation of God begins with creation (cf. Ps 19:1; Rom 1:20), opening the way for its scientific apprehension to serve as a theological datum in these reflections.

Theologians distinguish between "general revelation" and "special revelation." The phrase "general revelation" is used when referring to God's universal self-revelation, for God is present everywhere, while "special revelation" is reserved for God's manifestations in the history of Israel and specifically in the person and work of Jesus of Nazareth.

In contemporary society, Judeo-Christian claims to possess God's special revelation have become controversial. Opponents belittle the idea that the omnipresent God should reveal himself in a specific, localized, and historical manner because it runs against the grain of religious relativism; others raise questions about the credibility of biblical faith in the age of critical reason. These arguments often go together with another that accuses the Christian tradition of past pretensions to a triumphalism based on claims to special revelation.[9]

Criticisms such as these overlook the paradox inherent in the idea of revelation. If, on the one hand, the primary and most general meaning of revelation is God's self-gift to the world, then revelation is fundamentally universal and does not imply that groups or individuals to whom such revelation is given are superior to other people. At the same time, the notion of revelation is inseparable from a real human experience: people are encountering the mystery of the transcendent that inevitably surprises us with its newness and unpredictability. On the significance

---

7. Cited in Haught, *Revelation in History*, 14.
8. Haught, *Revelation in History*, 14.
9. Haught, *Revelation in History*, 16.

of this latter point, Haught observes: "If we lose the notion of revelation, we lose the sense that we are being addressed and invited to something beyond ourselves. And when we lose that impression of being challenged by the mystery of the transcendent, our world becomes closed in on itself in a way that is too suffocating for the human spirit."[10] Thus, an important dimension of human existence comes to light that, while within the reach of our perception and experience, lies beyond our control as it may break unexpectedly into our consciousness with incomprehensible newness, stimulating and challenging us to reimagine human existence along lines never imagined before.

In studies of religious experiences, such encounters have been referred to as hierophanies (sometimes theophanies). They highlight at least two hidden elements: the experience itself, which is accessible only to the person undergoing it, and the hiddenness of the mystery that discloses itself. This has led in the Judeo-Christian tradition to describing God's self-disclosure as an unveiling and as a veiling at the same time, since in the moment of disclosure a person may experience a special glimpse of ultimate reality while the mystery in its totality remains veiled.

Existentially speaking, revelation properly understood addresses us by engaging our need for transformation. Revelation is therefore always more than otherwise hidden information given to satisfy our curiosity. As a lure that urges us forward to embrace hitherto unknown realms of existence, revelation is always imbued with promise and hope. That's why, in the biblical tradition, where divine self-disclosure takes place through a word inscribed in human language, its truth claims are more promissory than propositional. While such words cannot be reduced to mere statements of factual correctness, let alone raised to the level of infallible dogma, they remain true words nonetheless. By the same token, we would commit a serious error if we understood by verbal revelation taking dictation from God. For when the biblical writers expressed what they experienced in an encounter with transcendent reality and what this reality seemed to have been "saying" when they encountered it, they employed their own language, culture, and state of knowledge. Such words work in us much like poetry that affects us intimately. When a poem "opens our eyes," reading it becomes a memorable event in our lives. But even in this instance we cannot escape the paradox that the words that reveal also veil. Not everyone who reads a poem will be touched by it

---

10. Haught, *Revelation in History*, 17.

experientially, nor must we assume that all beautiful words constitute revelatory truths, nor can we automatically interpret what we regard as beautiful, pleasing, or attractive as God's self-disclosure.

What's more, the Judeo-Christian tradition claims that revelation occurs in love and thus must occur in freedom. This implies that our experience must necessarily begin with what is *not* self-evident. After all, freedom would be lost if the experience were to convey something that is indubitably true, that could be neither doubted nor questioned. Therefore, as mentioned before, revelatory experiences cannot be like taking dictation from God. Such encounters simply invite us to receive a gift that is subject to human acceptance or rejection. Since it is impossible to receive anything without our consent, the phenomenon of revelation, like true love, must approach us in a noncompulsory manner, in which we are not even obligated to obey. Instead, revelation is offered—unlike erotic love, which is fundamentally self-indulgent—in the liminal space between unveiling and veiling.[11]

As we will see more fully in the next chapter, in Israel's case, ultimate reality is perceived in two modes: first, as a generic and abstract but awesome presence, which they called *Elohim*, and then as a personal presence—a presence of grace and love—who they believed had identified himself as *YHWH*. *Elohim* revealed himself in his acts, acting in and through his word, while *YHWH* reveals himself, also by speaking, as the personal rescuer and shepherd of Israel.

Christianity shares with the Jewish religion the ability to articulate itself in symbolic language, in detail as well as in its basic structure. Judaism is able to express its central divine revelation, the manifestation of God as *YHWH*, in genuinely symbolic narrations as the narrative texts of the Hebrew Bible show, just as the New Testament does in the parables. In short, the transcendent dimension of reality is manifested not so much in visual nature phenomena that are to be contemplated, like a sunset, but rather in series of occurrences in historical events.[12]

Finally, I want to draw attention to two often-overlooked features of a theology of revelation. In general, theology seeks to follow the path to truth, which is not an arbitrary endeavor and therefore must follow a method. The great Swiss theologian Karl Barth denied that theology can use "predetermined methods" when dealing with revelation because

---

11. Horner, "Words That Reveal."
12. Baudler, *God and Violence*, 29.

revelation occurs at the human level on God's initiative and therefore must remain independent of human inputs.

Others have rightly pointed out that if we start with revelation, how should we interpret it if our only reference point is nature? And how can we use nature as a proof of God's existence such that it becomes universally relevant? At the same time, we no longer start with nature as Thomas Aquinas did in the Middle Ages, for *nature as such*—that is, as an uninterpreted entity—does not exist. Nature for us is always already interpreted by the sciences, and the modern sciences have changed our perception of creation: space and time are now interpreted in terms of relativity; biology explains the evolution of life; the neurosciences, in clarifying our experience, map the operation of the brain, and so on. Conversely, of course, in our theologizing we no longer start with *revelation as such* either, but with revelation as interpreted by the church and theologians.[13] Additionally, considering contemporary scientific explanations, the God-creation connection is for many no longer self-evident, which leads us to ask whether seeing God engaged as sustainer of an ongoing creation offers a more meaningful option.

As important as it is to recognize Barth's concern, we cannot do without method completely, because for revelation to make sense to us it must respond coherently and relevantly to concrete questions in the contemporary context that the themes and subthemes of the book address: cosmos, the natural world, the human creature, the rise of human consciousness, the rise of monotheism, God in the age of science, God's glory in nature, and the inseparability of cosmos and Christ.

Revelation: The Cosmic Context

Prescientific cosmologies, beginning in the Paleolithic, sought to tell stories about all that exists. Although attempting to give a holistic account, such cosmologies even up to and beyond medieval times were thoroughly anthropocentric, featuring what the people saw and experienced, the concrete and the mythical. They acted like the early cartographers, who sought to depict the then-known world but marked the unknown with mythical references like "there be dragons."

Modern cosmology has dramatically and profoundly altered the human situation, and many of us have been troubled by the implications

---

13. Du Toit, "Towards a New Natural Theology."

of this development, especially by the question of how to live with intellectual integrity in this age of science while remaining faithful to the traditional teaching of the church about the creation. Today, most scientists agree that we live in an unfinished universe, one that came into being over 13.8 billion years ago in the initial irruption popularly known as the "big bang." Phenomena like star formation and star deaths are ongoing processes, like the forward urge of the universe toward higher forms of complexity and novelty. At the same time, there is less certainty about its future state and, from all we know, the present epoch seems to be only an early phase on the calendar of an even longer cosmic journey. (Needless to say, no scientist is prepared to ask, let alone answer, the question of whether the universe has any purpose.)

For the educated person living in the twenty-first century, this evolutionary cosmos has increasingly become their fundamental framework of existence, following the public descriptions of science. As the evolutionary cosmos is also the environment in which individual Christians are bidden to think and speak plausibly about their faith, it seems to me that this task has become impossible without having first gained a deeper understanding through reading and reflection of the claims of science and the theological questions this new cosmic horizon raises for Christianity.

What are (secular) questions that arise from the new cosmology? Does the universe have meaning? If so, what is it? Given the theory of relativity, does the universe have a history? What is the future of the cosmos? Is the universe life-friendly? Is the universe real or merely a simulation? Is our universe alone or one of many? For Christians different questions arise. If God is the creator of all there is, what does "revelation" mean in the cosmic context, especially considering its apparently evolutionary nature? What are the implications for the Christian (theological) imagination considering the vastness of the observable universe, including the possibility of life, even sentient life, elsewhere in the cosmos? Or, granting the cosmos's revelatory character, wouldn't it make sense to define its evolutionary nature, at the mundane level, as the fullest possible unfolding of the universe itself, and its ultimate stage as the consummation of the struggles of all the cosmic ages for a significance that might validate their arduous journey? (cf. Paul's intuition in Rom 8:19, 22). Haught observes: "From one point of view revelation is the surprising and interruptive utterance of a word of promise into what otherwise is interpretable as a cosmic void. But viewed from the side of the cosmos-in-evolution it is legitimate to see revelation as the flowering fulfillment of the universe itself. Revelation is, in one sense at least, the very purpose of the evolving universe."[14]

I do not intend to develop answers for questions like these but to raise Christian awareness that by closing our eyes to the new cosmic story as God's self-revelation Christianity is in danger of separating God's self-revelation in Christ from the creation as it is. I have become convinced that the task before us is nothing less than working toward a sufficiently expanded Christian imagination that leads to a God-honoring affirmation of unity between the two.

Building on the earlier vision of revelation as God's self-gift to the creation, one may now say that if God gives away the fullness of divinity to the cosmos, and the cosmos, because it is finite, cannot receive the immensity of God's self-gift in a single moment, it must do so incrementally.[15] Haught continues: "The cosmos moves and grows as a result of the implantation of the self-giving mystery that forever lies beyond it. Because of the cosmic self-transcendence 'time' is born. The meaning of time (which has always been a problem for philosophers) when seen in terms of God's self-revelation is that it is the mode of becoming that a world has to assume while it is receiving God into itself."[16]

Haught's vision implies that a now time-impregnated cosmos emerges as a world filled with promise. Unable to contain the infinite instantly, it must move forward in anticipation of the divine promise, which defines its future. And since it is out of the crucible of God's futurity that the revelatory promise springs forth, the cosmos is lured toward its fulfillment. Thus, cosmic evolution from the big bang until now and beyond, including the evolution of life on Earth and possibly elsewhere in the cosmos, becomes the story of creation's self-transcendence as it moves forward in time in response to God's promise. Science cannot see this. For the scientific observer, cosmic evolution is merely the process of gradual appearances of increasingly more complex organizations of matter, entities, and social structures, while "from the perspective of revelation cosmic evolution is the story of the God-of-the-future entering ever more intimately into the fabric of the universe."[17]

---

14. Haught, *Revelation in History*, 24–25.
15. Haught, *Revelation in History*, 25.
16. Haught, *Revelation in History*, 25.
17. Haught, *Revelation in* History, 5.

Revelation in History

All revelation occurs in history. Today, according to modern science, natural history or the history of the universe as a whole is measured in billions of years and the history of *Homo sapiens* in tens of thousands. Yet, many conservative Christians reject these figures, believing that they conflict with the account of creation "revealed" in Genesis. I address this issue more fully in the next chapter; here the focus is the historical nature of revelation.

If the whole of creation is revelation as mentioned above, then revelation has occurred not just in human history but from creation's first flaring forth, even though there were no eyes to behold or minds to seek out its meaning (although this raises the question of whether an unobserved event can still be termed "revelation").

Although cynics tend to dismiss the possibility of making sense of history as "more or less bunk,"[18] students of history are fascinated by the diverse patterns they find in the cumulative effect of events, human experiences, and actions. And so, the struggle continues to make history intelligible and meaningful at various levels—cosmic, human, religious. What has emerged as one of the most recent fields of interest is the study of "deep history," made famous by David Christian in his book *Maps of Time*.[19] This work unites natural history and human history in a single, grand, and intelligible narrative that draws on data from cosmology, biology, geology, and archaeology to locate our species in a larger cosmic context. This view of history underlies all that follows in this chapter and beyond.

Although animals live naturally in time, what sets us apart is our ability to experience time and so not only to become conscious of seasonal change, but to recall what happened in the past while our imagination enables us to make plans for tomorrow. In other words, ever since we became aware that we do not live in time like other creatures do, we have attempted to understand our place in the great chronology of life. Yet the familiar conception of time as linear is a relatively recent discovery.

By "history of revelation" we mean those revelatory events in the life of a religious community that have shaped and given meaning to its beliefs and practices over time. For the Christian community this history

---

18. Lockerby, "Henry Ford—Quote," https://www.science20.com/chatter_box/henry_ford_quote_history_bunk-79505.

19. Christian, *Maps of Time*.

begins with Abraham, culminates in the Christ event, and continues through the apostolic witness of the first century CE to this day. This history is expected to end with what has been metaphorically called the kingdom or reign of God, of which Jesus of Nazareth is believed to be the proleptic manifestation in history.

The most reflective of the early Christian historiographers is undoubtedly Augustine (354–430). For him, God's eternal plan was unfolding in a time-ordered fashion on a straight line from the fall of Adam and Eve to the birth, death, and resurrection of Jesus as the history of the creation moves toward the final judgment with infallible predictability. Augustine's *City of God* (413–426) describes the lives and nations on their long redemptive path from original sin to the revelation of Jesus Christ. According to Augustine, history is characterized as the struggle between the elect—the inhabitants of the City of God—and the unbelieving world, or the City of Men, in its self-assertion against God. Thus, "sacred history" as an overarching narrative gave meaning to human existence, either as a statement of hope in a redeemed future for those who believe or as a tragedy for others. This view that interpreted history in terms of a linear series of inviolable events enjoyed canonical status in Europe throughout medieval times and continued to influence the interpretation and philosophy of history until the nineteenth century. For the church, this view still provides, in broad outline, the contours of redemptive history.

The Jewish community believes that God entered human history to accompany his people on their journey. The Hebrew Bible is not a philosophical construction or a system of knowledge but consists of a series of stories that are not myths or deliberations that reveal or hide "objective and abstract truths," but *one story*. It is the history of the people of God and their wanderings, their history of loyalty and disobedience, in which the nature of God in dialogue with his people is revealed.[20]

Until the biblical revelation, people did not expect love from God. Indeed, for most of human history, people have been afraid of transcendent reality and its appearance in history (however they might have called it). Prescientific societies perceived the divine with a sense of dread: when God appeared, someone was going to die, some punishment was going to be meted out, or a price of appeasement was going to be exacted. In short, humanity, for the most part, did not expect to be loved by God. When we consider that religions started out as nature religions, this view may

---

20. Ellul, *On Freedom, Love, and Power*, 23–24.

not be too surprising as they vested the often-destructive forces of nature with transcendent reality and power.

When speaking of revelation today, we no longer start with nature, but with human testimony, oral or textual, whereby the church gives primacy to the latter.[21] Hence the struggle of the early Christians with the issue of including Jesus in the divine identity. After all, this meant including a man in the divine identity as the Hebrew tradition understood it, a man who had lived a fully human life, then suffered rejection and the shameful fate of crucifixion. The inclusion of these elements could not leave their understanding of *YHWH*, their God, unaffected. The astounding fact is that the New Testament writers (most of them Jews) concluded that the identity of God now needed to include the history of Jesus, yet they continued to see in this redefinition the God of Israel. Much of this adaptation has to do with the human ability to understand symbols. As we will see in a later chapter, it is this ability to experience reality symbolically that lies at the heart of the dialogue between religion and culture.[22]

To round off our reflections on revelation in history, a word about an often-unrecognized yet critical aspect of historical existence is needed, namely, the indeterminacy and open-endedness of the future. As mentioned, just as the cosmos is unable to contain the infinite self-gift of God in one moment and so must move forward in the hope of future fulfillment, so must human beings embrace the tension between their finiteness in the face of the unconditional and infinite love of God. In other words, when we undergo a measure of God's self-disclosure, we too are challenged and lured to embrace hitherto unknown possibilities of existence. Because of the promissory nature of such events, we will experience a foretaste of what is yet to come, but we must also wait for the fulfillment of an open-ended future, which, in the Christian tradition, means ultimate union with God.

What beckoned our biblical ancestors to move deftly and trustingly out of natural existence into historical existence, and with it into an unknown future, was the divine promise underwritten by nothing more than the assurance of God's faithfulness. Since it is the divine promise that defines the future, it is a future that springs forth from God's own futurity, out of which the evolution of the cosmos moves forward in hope.

---

21. The believing community validates oral testimony of revelatory experiences in light of the biblical text.

22. Baudler, *God and Violence*, 11.

In the Christian conception of the world, the divine promise is the answer to the unbearable weight of the uncertainty of an otherwise open future and to the depressing prospect of cosmic loneliness. Thus, "the quest for revelation . . . may be understood, in the present context at least, as the quest for some resolution of the mistiness that confronts us as we peer into the unknown outcome of historical events."[23] This kind of faith raises an important question, though. Is there any evidence from past patterns that would allow us to draw meaning from history, a possibility that most modern intellectuals would deny? Yet, despite the denial of a meaningful future, the hunger for a deeper meaning of history and for a better world has not died.

The good news is that biblical revelation, inseparable from the notion of promise, will not give up the hope for a meaningful future of historical existence as it looks to the crucified God in Jesus Christ and his resurrection as its central symbol and guarantee of the promise for a new creation.

Personal and Public Revelation

C. S. Lewis (1898–1963), one of the most influential Christian writers of his day, a fellow and tutor in English literature at Oxford University until 1954 and later the chair of medieval and Renaissance literature at Cambridge University, reached a vast audience as a Christian apologist. *The Chronicles of Narnia* alone have sold over 100 million copies. Though he started out as a militant atheist, his worldview was transformed into faith in God, a process that he describes as very intellectual and very gradual. In his childhood, Lewis seems to have experienced from time to time a "sense of intense longing" beyond his desire for connection with something or someone. Later he began to realize that this longing could never be met through any human relationship, perceiving it as an experience that pointed him clearly to the Creator. In 1930, he became a Christian believer in the Anglican tradition. His book *Surprised by Joy* (1955) recounts his spiritual journey during his late twenties and early thirties.[24] In our context, we would consider Lewis's conversion the result of a personal revelation. Since his faith emerged gradually over time, the underlying illumination was not limited to a single moment but occurred

---

23. Haught, *Revelation in History*, 35.
24. Lewis, *Surprised by Joy*.

process-like, in stages. In other words, at one point in his personal history Lewis felt he had moved from unbelief and rejection of God to faith and trust in him, which in turn led to his subsequent acceptance of the Christian scriptures and doctrines as reasonably credible.

The conundrum of personal revelation is this: what had become so obvious to Lewis that he staked his intellectual and personal life on it was *not* obvious to everybody! In other words, personal revelations, as subjective claims for which the only evidence is the word of the person who advances the claim, can only be believed but not known. Identical problems arise with worldviews based on claims of personal revelation. Today, millions of Muslims believe that Jews and Christians will burn eternally in hell because they don't believe the revelation given to Mohammed in the Qu'ran, while millions of conservative Christians believe that Muslims face the same predicament because they do not believe as they do. The result? Countless religious conflicts and persecutions that have disillusioned many. Today, as in the past, such conflicts have taken on geopolitical dimensions in the so-called war on terror, while the intellectual world is divided over the ontological question of whether a transcendent intelligence exists beyond and behind the universe.

Our ancestors perceived such a transcendent presence behind all that was. What allowed them to make sense of the world and of their place in it was their ability to express meaning, first in symbols, later in words and metaphors. They saw divinity in bluebells and waterfalls, in the regularity of the seasons, in the migration of birds and buffalo, in the stellar movements of the night sky, in the miracle of birth, and in the mystery of death. Theirs was a natural theology that spoke to them of a wholeness beyond themselves to which they belonged nonetheless, of something or someone that encompassed all that they beheld with their eyes and perceived through their senses. As they stood, often amazed, before this mystery that provided sustenance and gave cohesion to their world, they somehow felt addressed by it. They gave it names and so formed in response a tangible as well as symbolic conception of the world, integrating transcendent reality with the experience of life. This was their way of knowing that never questioned the reality of this *mysterium tremendum et fascinans*, as Rudolf Otto perceptively named his concept of the numinous. The mystery just was. A world without it was unthinkable.[25]

---

25. Otto, *Idea of the Holy*, ch. 4.

Such an understanding of the world provided the background for an oral tradition from which eventually the sacred texts should emerge. The personal revelations of prophets and psalmists, seers and sages, and lastly apostles, received in visions and dreams or in encounters with the mystery, were shared with others in the group or tribe, thus bracing their identity. Through retelling, debate, and agreement, these stories took on infallible and authoritative qualities and in their later written form became sacred texts. With rising monotheism in Israel, their texts were believed to have been "God-breathed," expressing not only Israel's authoritative history but also God's direct commandments, prohibitions, and promises to be taken as rules of life by God's special covenant community. In other words, the revelations contained in the Hebrew Bible did not fall from heaven as a single event; instead they emerged. They were redacted and refined as the result of a long process lasting hundreds of years until the text reached its present form.

Similar observations may be made for the New Testament. While initially products of personal revelation obtained by individuals (prophets, apostles), these canonical texts are taken today as true and authoritative, and their words now normatively guide the life of Christian communities and traditions everywhere. What were once personal revelations were thus transformed into public revelations by a process of traditioning. Their validity rests today not on proofs in a scientific sense but on extensive communal consent. While different traditions have come up with different emphases, all Christians believe that God's revelation was completed and perfected in Christ, the final mediator, author, and interpreter of all revelation.

But what of personal revelations today? Although they may be recognized in various ways, it is generally agreed that they are not allowed to surpass or correct public revelations, which are binding on all believers. In the Catholic tradition, the phrase "public revelation" also includes the views of the Magisterium, the sacraments, and the Catholic dogma. Therefore, when clarifying meaning, the work of theologians is considered indispensable.

In this context, we must address one further question before we can move on, the question of religious freedom, among other reasons because it underpins the independence of science from religious influence and thus the emergence of modern secularity. The concept of religious freedom has its roots in the Protestant Reformation. The premier example of this stance is Martin Luther's refusal to recant his views before the Diet of the Empire in 1521. Faith, according to this disposition, was the free work of the Holy Spirit in the believer; therefore it could not be forced on a person. This view was later used to promote the separation of church and state as well as the principles of tolerance, which became the basis of many democratic constitutions as well as of the legal-philosophical thought that shaped such public instruments as the American Bill of Rights, the Declaration of Independence, the Universal Declaration of Human Rights, and more. Another development that issued from the quest for religious freedom was the emancipation of the sciences from religious interference, although the split between science and religion occurred a century earlier.

The Problem of God's Hiddenness

In an earlier section, we touched on the idea that God, despite the magnificence of his creation, remains hidden based on the explanation that we could not bear the full weight of God's total self-disclosure. This raises an important question for us, because when we, as creatures, speak of revelation (and even claim to possess a capacity to resonate with it), we must also confess that when it comes to knowledge of the divine, we do not stand on our own ground. After all, neither our natural insights nor philosophy or science can say anything definitive about God or about revelation, not even whether there is such a phenomenon. At the same time, from our earliest beginnings we have oriented our gaze upward in the hope of comprehending the mystery that addresses us. Even when we engage in some form of natural theology, we cannot assert with any sense of certainty that knowledge of God will follow. All we can do is say what God is not. Where then shall God's revelation be sought? Isn't nature, as creation, a primary manifestation of God's action? Therefore, it should be possible to deduce that there can be no form of existence that does not point to God and no space where God is not present. So, we may say that God's presence is unmistakable but undefinable. Hence, it can only be analogically understood because of the great distance between the transcendent God and the natural world. The Bible, too, alerts us to the fact that God actively hides himself from his human creation.[26] What, then, shall we make of the problem of God's hiddenness?

---

26. Isa 45:15; Matt 11:25; Rom 11:33. See also the Song of Songs, where the lover hides from the beloved until she yearns for him.

THE NATURE OF REVELATION

In the parable of "The King and the Maiden," the Danish philosopher Søren Kierkegaard narrates his explanation poetically. If God appeared in divine form, it would be more than mere humans could handle. So, when the king wants to express his love to the maiden, he approaches her not in royal splendor, out of concern that it might overwhelm her and that she might respond from motives other than true love. Instead, he approaches her at her level, dressed in a servant's cloak, revealing the true nature of his intentions as well as respecting her as a person and acknowledging the equality of true love.

Sensitive to Kierkegaard's insight, one theologian has perceptively commented that divine revelation must necessarily contain an ambiguity. On the one hand, it must be readily accessible; on the other, it must also be easily resistible. Only in this way can it ensure that faith remains grounded in personal encounter while engaging the believer's conscious participation: "God would make knowledge of himself widely available for those who wish to have it, but God would not wish to force such knowledge on those who do not wish to know God."[27]

Yet, the New Testament and the Christian confession draw the curtain of hiddenness aside: Jesus' identity is unmistakably defined as "the image of the invisible God" (Col 1:5). Statements such as these express in another way the thought and teaching of Jesus, who declared, according to John 14:9, "He who has seen me has seen the Father . . . " What follows in the ensuing chapters is rooted in this confession. While Paul affirms—and what the church believes—that "in him the fullness of deity dwelled bodily" (Col 1:19), the church in this age remains set on the footing of faith and the yet-to-be-fulfilled yearning for a deeper experience of God's glory.

Revelation or Science?

Several decades ago, renowned physical biochemist and theologian Arthur Peacocke (1924–2006) warned that in the age of science Christianity would face a credibility crisis unless the churches took scientific discoveries seriously and incorporated their implications into Christian theology. He argued that the alarming dissonance between church language and the way people perceived the world disaffected ever larger numbers of people, while an underequipped leadership (ordained and lay) was

---

27. Evans, *Natural Signs*, 12–17.

unable to meet the adaptive challenge.[28] Studies of religious attitudes in Western Europe have confirmed Peacocke's prediction. For example, according to a survey conducted by the Pew Research Center of 25,000 adults across 15 countries, the Christian revelation is losing traction in Western Europe on a large scale. The heartland of the Protestant Reformation and the place where Catholicism had its base for most of its history has become one of the world's most secular regions.[29] When Notre Dame Cathedral caught fire in 2019, the *Washington Post* drew attention to dwindling church attendance: "In France and across Europe, pews are emptying, seminary classes are shrinking, and some parishes are borrowing priests from other continents as a way of staying afloat."[30]

While most of the participants in the Pew survey said they were baptized (91 percent), claimed to have been raised as Christians (81 percent), and currently consider themselves Christians (71 percent), many reported that they resonate with some "other spiritual force" or with God-denying statements like "There is no higher power in the universe, only the laws of nature" and "Science makes religion unnecessary in my life." In short, religious focus and practice in Europe have shifted away from the Christian revelation toward secular humanism. While more than two-thirds still value their Christian identity although they may not attend church, almost half (46 percent) no longer believe in the biblical God. In addition, the portion of active Christians (who attend church at least once a month) has shrunk to less than one-fifth (18 percent). Those with no religious affiliation have increased to nearly a quarter (24 percent), while those who answered "Don't know" or "Other" make up the rest (5 percent).[31]

---

28. Peacocke, *Palace of Glory*, 4.

29. Pew Research Center, *Being Christian*, https://www.pewforum.org/2018/05/29/being-christian-in-western-europe/.

30. Harlan, "Fire Came," https://www.washingtonpost.com/world/europe/notre-dame-fire-came-at-a difficult-time for-french-catholics/20119/04/04/. This was in response to the fire that destroyed Notre Dame Cathedral.

31. This picture may even be optimistic. Researchers have long suspected a gap between reported church attendance and actual practice. Many respondents view survey questions about church attendance as questions about their identity rather than about their practice and so tend to inflate self-reported data. This is not because they want to be dishonest, but to protect their self-image as good Christians based on unrecognized psychological grounds. "Church Attendance," https://en.wikipedia.org/wiki/Church_attendance/.

What is surprising in the survey is the strong influence of science. In most countries roughly a quarter or more of nominal Christians say that science makes religion unnecessary to them, a view even held by a substantial minority of practicing Christians. The survey revealed among other things that Europeans increasingly prioritize science over revelation, which explains why in regions where science predominates and where people are more educated Christianity is in decline. In short, we are witnessing what Arthur Peacocke saw coming in the 1960s, namely, a growing weakness of the churches' mission to Western humanity as a result of failing to find, in the age of science, convincing ways of expressing their beliefs. When their theologians failed to take scientific discoveries seriously, people in the pew were left without the intellectual tools that would have enabled them not merely to cope with but to outthink the contemporary cultural and intellectual changes. The result: declining membership.

Speaking from personal experience as a Christian for almost 50 years, regularly attending church in a variety of traditions, I have yet to hear a church leader preach on the God-glorifying value of scientific discoveries, especially the revolutionary ones that no longer fit neatly into Christianity's traditional view of creation. These include the discovery that the Earth is not flat, resting on pillars surrounded by primordial waters, but a sphere orbiting the sun; the discovery of the telescope and the microscope, which changed our ability to investigate God's creation forever; the discoveries of cosmology that we live in an expanding and ongoingly creative—that is, unfinished—universe; the discovery of electromagnetism, of aerodynamics, of organic chemistry, of antibiotics and other health- and life-promoting medicines; the discovery that biological and genetic change takes place over time and is called evolution; the discovery of quantum physics, and so much more. Today, just as physics no longer makes sense without gravity and Einstein's general relativity, so biology does not make sense without evolution.

All these discoveries confront thoughtful Christians with an unavoidable question: What did God have to do with them? Did God—who gives existence to every segment of time—have anything to do with the rise of the scientific age, with the creative imagination of the scientists who developed the rigorous methods of investigation that have led to these extraordinary unveilings of nature's secrets? Did God have anything to do with the sophisticated technologies that enabled us to land on the moon, to set up a laboratory on Mars, to probe the rings of

Saturn, and to transmit the results over billions of kilometers back to Earth? Honestly, what was God's role in all this? Creating an intelligible universe? Creating human beings with gifts that would progressively discover this intelligibility and describe it with increasing precision? If the answer is yes to these questions alone, wouldn't it be reasonable to take these unveilings seriously, even regarding them as "revelations" of the public kind intended for the good of humanity and of creation at large at this stage in history? And further, couldn't we argue that if the church is largely ignored today, it is not because science has triumphed over religion but because science has so radically reoriented the cosmological and anthropological platforms of culture that the biblical perspective makes little sense any longer, unless expressed in new ways? If this is correct, wouldn't it be reasonable to say to the churches that they must tackle this task with urgency or risk losing contact with modern culture to the point of becoming irrelevant? To move this conversation forward, I propose a generalized link between the results of science and the notion of divine revelation in the next section.

Science as Revelation?

In the sixteenth century, based on a fresh interpretation of already existing astronomical data, Nicolaus Copernicus (1473–1543) proposed the hypothesis that the Earth orbits the sun. Since the observations of his database were repetitive enough, astronomers and mathematicians of his day regarded them as "facts," meaning true in the language of science. The problem was that his proposal of a sun-centered planetary system departed so radically from the traditional scientific view of the universe and from church doctrine that Copernicus decided to postpone the publication of his work until his deathbed. Today, 500 years later, we know beyond a shadow of a doubt that the Earth orbits the sun. For us, it is no longer a matter of believing or hypothesizing, but of fact. Indeed, today we possess more reliable evidence for the correctness of this claim than Copernicus ever had, along with deep public acceptance. Our database includes evidence from countless observations as well as pictures taken in real time on numerous spaceflights to the moon, to Mars, to the moons of Jupiter, and more. What once was a dangerous hypothesis has become a public truth.

A similar case may be made in relation to another, more recent "dangerous hypothesis." In the mid-nineteenth century, Charles Darwin (1809–1882) went public with his proposal that the complexity and diversity of life on Earth had emerged gradually from earlier, simpler life forms over vast spans of time through the mechanism of mutations and natural selection. Like Copernicus, Darwin was aware of the pioneering nature of his proposal and afraid of public reaction, so he postponed its publication for 20 years. Since then, this discovery has scandalized many and even today acceptance of evolution is far from universal. Although thinking in evolutionary terms is widely accepted by most scientists, it continues to be rejected by large numbers of people. They do not just oppose the idea but also the massive body of scientific evidence in its favor.[32]

About one in five adults in the United States and millions of Christians around the world believe that life on Earth did not evolve at all but was created as is. While it is correct that science does not yet fully understand what drives biological change, the observational evidence that the history of biological change over time from simpler life forms to highly complex ones is so vast that it must by now be considered a public fact. One can go as far as to say that evolution is so deeply embedded in biology and associated disciplines that eliminating it from biology would be like eliminating gravity or general relativity from physics and cosmology.[33] Quite simply, we have no better alternative than evolution to describe the history of biological change.

Certainly, several factors have contributed to this idiosyncrasy. One is the incorrect use of scientific language in everyday contexts. Let me take an obvious example. In everyday usage it is common to speak of a "theory" when referring to a half-digested idea without much supporting evidence. We say, "Oh, this or that is just a theory." Not so in science. Here we begin with a "proposal" or "hypothesis." It guides experimental tests,

---

32. It is ironic that in the courts forensic DNA testing has been widely accepted as standard practice on which juries and judges rely for their judgments. This practice turns on evolutionary changes recorded in the DNA, yet millions of Christians still doubt the validity of evolution. Another irony: millions of Christians travel by air every year, literally entrusting their lives to the science of aerodynamics, which makes air travel possible. Yet they reject the findings of biological research, which employs the same scientific method for its conclusions.

33. Nearly all biologists acknowledge evolution *as a fact*. While the term *theory* may still be appropriate to models that explain the origin of life or how life evolves, the term has become redundant regarding the question of whether life evolves.

which in turn may produce repeatable data; when the data eventually confirm the hypothesis, we may speak, perhaps after some modification, of a "theory." Yet theories are not facts in a strict sense. Although we may treat them as such for practical purposes, we must observe an important distinction. Since in science all knowledge is provisional, even established theories must submit to (new) data. In general, however, theories in science are well-informed structures of ideas of strong interpretive power regarding the existing dataset. They are firm enough in the presence of empirical evidence to be treated as facts until modified or falsified by fresh evidence.

Extending this thought to theology, I submit that if we believe in God as creator of the natural world, then scientifically discovered facts about the inner workings of his creation are inherently part of God's creation. For instance, the fact that the Earth revolves around its axis in cycles of 24 hours and orbits the sun once a year belongs to God's creation, just as the measure of its circumference, its mass, its distance from the sun, and so on do. On this logic, the discovery of these parameters constitutes an act of revelation as they tell us something about God's creation not known before. Since all truth is God's truth, the worldwide community of scientists, as they honestly and collaboratively search out facts of this nature and submit their work to critical reviews by their peers, are engaged, whether they know it or not, in acts of revelation. Moreover, what must be counted as part of God's creativity in creation is the emergence of the scientific age with all its complexity. After all, this age has emerged and developed together with the abilities necessary for such discoveries. We must further assume that these discoveries were intended to be public knowledge. If so, these discoveries belong to the realm of public revelation, and what the scientific community discovers are ciphers through which God speaks to us at this point in history for the common good. But there is more.

Scientists in the process of unveiling what has been hidden in previous ages help us to enlarge our vision of God as creator and sustainer of a cosmos far grander and more complex than we could ever have imagined. Hence, we must grasp that in all the creative processes at work continuously from the big bang to the present God has always already been there in their randomness and regularities long before scientists observed and described them. To call the Earth flat when we know that it is spherical or to deny biological change over time when it is measurably evident amounts to, in the age of science, speaking falsely about the work of God in creation, or worse, cleaving the inseparable wholeness of God's self-revelation in creation and in Christ.

Conclusion

In our time, science and technology have profoundly altered the cosmic and societal perceptions of the world. At the same time, large parts of Christianity have rejected scientific evidence that seemingly conflicts with traditional views. This constellation makes it necessary to think afresh about how thoughtful Christians can speak with intellectual integrity about God's creation. This is the rationale for our exploration in the preceding sections. Beginning with the reality of an all-encompassing, even ultimate mystery, we noted that we encounter this mystery in the midst of the beauty, complexity, and precariousness of human existence. It addresses us in our living and dying and in such encounters we become aware of our need for transformation or self-transcendence. The common human response to mystery is religion.

From a Christian viewpoint we saw how the creation, as described by modern science, especially by cosmology, invites us to enlarge our vision of God far beyond current and traditional conceptions. Explaining the meaning of revelation as an unveiling, we noted how such a simple idea was nonetheless fraught with difficulties. There is the charge that revelation, which is usually personal, is too subjective for the age of science; or the question of origin: is it divine or merely the result of advanced brain functions? Anyhow, passing through the human mind, it remains deeply conditioned by nature.

Then there is the overarching question on which the entire edifice of Christianity rests and that in turn rests on faith: the question of God's self-revelation in history. Here we noted with Karl Rahner that God's revelation is "the communication of the mystery of God to the world" that influences the world "in every phase of its coming into being," so that the cosmic dimension becomes evident. John Haught's definition of God's self-revelation as "creative, formative love" loops revelation back to the very beginning, the moment of creation. We further noted that humans, as a "cosmic species," have tried from Paleolithic times to make sense of what they experienced—the concrete and the mythical—by telling stories about it. Stories provided meaning and community cohesion.

However, modern cosmology has so dramatically altered the human situation that many people feel disconnected from it. For one, the cosmos has become vast while Earth has shrunk to infinitesimal smallness, which raises new existential questions for humanity: how to live in a world that is unfinished with a future that is open in a cosmological sense. In this milieu, the task of the Christians community to engage meaningfully with the new cosmic story could not be more urgent. How else will it substantiate its claim that it possesses the revelation of the "cosmic Christ" as the self-revelation of the Creator?

2

The Biblical Account of Creation
Interpretation and Meaning

THIS CHAPTER REEXAMINES THE creation accounts in Genesis 1 and 2 and their implications for the creation-evolution controversy. My purpose is twofold. First, I wish to offer help to those who struggle with the conflict between their commitment to the biblical text and their desire to bring the discoveries of science into their theological thinking for the sake of intellectual integrity. My method will center on a Hebrew reading of these accounts of origin that never claimed to tell us how creation happened. Second, I wish to discharge a huge debt of gratitude. I owe it to two renowned scholars, Joel M. Hoffman, a Hebrew linguist and translation theorist, and Jacques Ellul, one of the most important Christian social and theological thinkers of the twentieth century. The former drew my attention to the common fallacy that most readers of the Bible assume that they know what it says because they have read it in English, overlooking that our translations from the Hebrew contain substantial mistakes and significant errors. Even well-loved quotations like "The Lord is my shepherd . . . ," "In the beginning, God created the heavens and the earth . . . ," "Thou shall not covet . . . ," and "Let my people go . . . " are not accurate reflections of the original.[1] I am indebted to Jacques Ellul for opening my eyes to the struggle involved in the quest that seeks

---

1. Hoffman, *And God Said*, 3.

to honor the revelation of God based on the ancient texts while living in the contemporary world. I rely on him here for nuanced interpretations based on the Hebrew text and Aramaic sources.

Text and Revelation

After a journey of 68 hours, the crew of *Apollo* 8 orbited the moon 10 times; during their last orbit, on Christmas Eve 1968, they read the first 10 verses of the book of Genesis. Televised, it became the most watched TV event ever, as if to remind all peoples that the cosmos we inhabit has an absolute beginning.

Millions of Christians hold to the belief that in the opening chapters of the Bible one can read a step-by-step report of how the universe was created. In this chapter I show on evidence from the text itself that such a reading has never been the objective of Genesis 1 and 2, for Hebrew learning is more concerned with potential and possibilities than with final answers.

This claim may surprise, even scandalize, some of my readers, while others may breathe a sigh of relief, as it releases them from an almost unbearable inner conflict as they seek to be faithful to the Scriptures while living with intellectual integrity in the age of science. After all, God does not reveal by means of systematic knowledge but enters human history and accompanies his people. That is why the Hebrew Bible is not a set of abstract (scientific) truths—in this case about the creation—but a series of stories of the history of God and his people. It is the history of God's loyalty and of Israel's disobedience, and thus a historical testimony that God is for us.

Let me say at the outset that the Christian doctrine of creation is based on divine revelation and can be understood only from the standpoint of faith (Heb 11:3). God's work in creation is no less hidden from human understanding than the mystery of redemption; both can only be perceived by faith. In other words, when I propose to offer a fresh reading and thereby show that the text does not intend to offer a step-by-step account of creation, I am not advocating that we set the Genesis text aside. Instead, I am affirming the text as the bearer of the revelation of the God who reveals himself in a vertical divine-human relation, yet at different levels of meaning involving sounds, words, phrases, concepts, and actions.²

The biblical approach is thus clearly distinguished from the scientific. This does not mean that the latter is without theological implications, considering the enormous shifts science has wrought for our understanding of the cosmos, its beginning, and its dynamic conditions. For that reason alone, the Genesis accounts of creation deserve a closer look.

With the foregoing in mind, we must be prepared to face several interpretive problems, especially regarding certain meanings that the traditional view ignored as it read the creation accounts as reportage of what took place. As I probe the text's revelatory intention these issues will be addressed under a set of subheadings. At the end, I summarize the results, hoping to have succeeded in showing that the biblical text neither contradicts nor conflicts with a scientific understanding of the cosmos.

As we proceed, I draw attention to an unavoidable cultural conditioning our customary reading undergoes. After all, it takes place in our own language and cultural context. Readers therefore relate the text unwittingly to inherited cultural values, including religious ones, and to everything else they know. In addition, the text has come to us translated from two ancient languages, Hebrew and Greek, both very distant from our own categories and concepts that offer us meaning. Therefore, any accommodation to our own culture-based knowledge, symbols, and values can cause the text to say what it was never intended to say in the original. It goes without saying that this kind of cultural drift can only lead to interpretations based on preconceived ideas and to claims that differ from what the text was originally intended to signify. Apart from these precautions, we must approach the biblical text with the intellectual humility it deserves. One further preliminary point needs to be made. Interpreting key phrases of the Hebrew text, with Ellul's help, enables us to disentangle the first two chapters of Genesis from any baggage of religion and morality that we so easily bring to it.³ Importantly, the nuances we encounter have been there all along. Sadly, most of us have been too well trained in the traditional view to notice them any longer. When we allow these meanings to come alive for us, we are bound to discover

2. Hoffman, *And God Said*, 76.

3. This chapter relies on a series of lectures by Jacques Ellul masterfully compiled by one of his former postdoctoral students, now Professor Emeritus Willem H. Vanderburg. See Ellul, *On Freedom, Love, and Power*. For a better understanding of Ellul's reading of the Bible, the reader may find it helpful to consult Ellul's "Innocent Notes."

to our surprise how open these ancient texts are to an understanding of God's creation that neither precludes nor conflicts with the findings of modern science. By the same token, I am not suggesting that readers accommodate themselves to "the flat, thin world of our dominant culture," as Walter Brueggemann cautioned, for it can never be "an adequate venue for the abundant life given by the God of the gospel."[4]

Approaching the Text

To a surprising extent, the first two chapters of the Bible parallel the new cosmic story the sciences tell. The Old Testament (OT) presents a vision of the universe that is both dynamic and historical. If this text is about anything, it is about the possibilities and promises that the God of Israel was believed to have built into the creation. Its story, narrated by prophets and chronicled by sages, is complete with conflicts, chance events, and contingencies, yet filled with an ever-present openness that invites exploration and discovery. Scientists have found similar characteristics in their discoveries that underlie the new cosmic story. It is full of "symmetry breakings," new orders emerging from chaos, yet also discoverable lawfulness that describes a universe in which both creativity and disorder seem to vie for supremacy.[5] For the authors of the biblical creation accounts, the source of stability and order was the faithfulness and creative power of the Creator. It gave the universe its lawfulness, which modern science discovers and narrates in precise mathematical equations.

Today, those who plumb the depths of physical reality, be it the quantum nature of the subatomic realm, the molecular intricacies of living organisms, or the complexity of the mind-body interface, gain the impression that an all-encompassing wisdom seems to pervade the universe. While scientists would overstep their mandate if they claimed to possess knowledge beyond their observations and theories, it does not stop them from posing questions of ultimate concern. But it is here that science finds its limit, as famous cosmologist Stephen Hawking showed

---

4. Brueggemann, *Introduction*, xiii.

5. Symmetry breaking, a feature of pattern formation, is a phenomenon in physics, chemistry, or biology where a system in a symmetrical state ends up asymmetrical after some disturbance. In the early universe, pressure and temperature were so high that they prevented the formation of permanent elementary particles. When the universe cooled, the symmetry broke. As a result, mass and space-time separated, while the energy that was released enabled the formation of elementary particles.

when he asked: "What breathes fire into the equations and makes a universe for them to describe?"[6]

Biblically, the universe has always been understood in the context of human life and humans as creatures among other creatures that depended on the Creator. With the creation, time had come into existence, so that the universe and all its creatures past and present had a history. In addition, the arrow of time was pointing irreversibly toward a future that is ultimately subject to the Creator's noncoercive ordering in love. In the Hebrew tradition, sanctifying time was an important element. It saw every moment as unique and precious, which is climactically reflected in their Sabbath celebrations. Even servants and beasts of burden deserved their rest day once a week. Taken together, this historical perspective precluded the idea of endless repetition, so prevalent in the surrounding cultures of the time. For the Hebrew mind, things were perishable, but time was not, rendering every moment evanescent and filled with innovative potential inherent in the act of creation.

With these thoughts in mind, let us look at these two creation accounts more closely, aiming to understand the ancient texts on their own merits rather than through the eyes of the controversy that has mired their interpretation for so long. At the same time, we must not pass over this sad chapter in silence. After all, the acrimonious debate that the teaching of young-Earth creationism has ignited in American schools in its fight against evolution is still at an all-time high.[7] Sadly, it casts unjustified aspersions on Christianity as being unscientific, even antiscientific, a perception now vociferously exploited by the New Atheists.[8] Those who use the first two chapters of the Bible to justify their absolute conviction that the world was created in six 24-hour days only demonstrate their unwillingness to critically examine their own presuppositions, for, simply put, the first two chapters of the Bible are not about such matters.

Significantly, the Bible presents us with two creation accounts, the first in Genesis 1:1—2:1–3, and the second in Genesis 2:4–25. I will refer

---

6. Hawking, *Brief History of Time*, 174.

7. While all who believe that God created the universe are "creationists," young-Earth creationists believe that God created the world in six 24-hour days less than 10 thousand years ago more or less as is, including all the animals.

8. The term *New Atheists* refers to a group of thinkers and writers who promote the view that religions, especially Christianity, are superstitions based on irrational assumptions. Their irrationality should be exposed and their influence removed from the public square. The most prominent proponents include Richard Dawkins, Daniel Dennett, Sam Harris, and the late Christopher Hitchens.

to them in those terms. By reflecting on their meaning, I hope to show that these ancient texts are quite compatible with a dynamic, contingent, and even evolutionary universe, and that these characteristics are more consistent with a theology of self-giving love than a finished creation and a static cosmos implied by the traditional reading. Yet my interpretation is not a naive attempt to conflate the biblical creation accounts with models of modern cosmology. Rather, I attempt in the subsections that follow to delve deeper, with Ellul's help, into crucial biblical themes for better understanding of the Hebrew text. But first a word about "biblical inspiration."

Inspiration

Evangelical Christians believe in "plenary verbal inspiration," meaning that God's inspiration of the biblical authors extends to the very choice of the words they used. Each word is exactly as God intended it. The writing of the Scriptures was not a matter of mere guidance, but something given or imparted to them. Thus, the authors have sometimes been called "God's penmen." In short, when conservative Christians use the term *inspiration*, it is understood to be verbal and plenary—that is, every word, as come from God, is true, normative, and authoritative. However, there are a few problems with plenary verbal inspiration. First, those who assert that verbal and plenary inspiration does not mean "dictation" cannot clearly say how. It has been suggested that the Holy Spirit "directed the thoughts" of the writers, influencing their understanding in such a way that they would choose one word over another or to write "precisely the thoughts that [God] wished to express."[9] However, this argument merely relocates the problem to the subconscious level. There are other difficulties with the claim that every word is inspired. First, no Jewish interpreter has ever held such a view. Then there are questions of whether inspiration extends to incidents when the Bible quotes external (noncanonical) sources, even pagan authors. Or to quotations of OT texts in the New Testament (NT) where the authors don't seem too concerned about exact citations, but rather with conveying meaning. Or to the frequent grammatical errors we find in the book of Revelation. Then there is the problem that this claim applies only to the original autographs, which we do not possess, but not to later copies and translations. Lastly, the notion of

---

9. Erickson, *Christian Theology*, 215–16.

plenary verbal inspiration cannot adequately account for matters raised by historical textual criticism, textual differences, and the compositional construction of some biblical texts.

On a broader canvass, the notion of word-for-word inspiration not only runs against Jewish understanding of the text but has also led many Christians to go as far as asserting that the Bible is "inerrant" in all matters of biology, cosmology, geology, history, physics, psychology, and so on. Others believe that it is "inerrant" only in respect of its primary purpose, to reveal God's character, God's judgment and saving purposes, and God's good news to humanity. These tight schemes of biblical interpretation have, according to Ellul, led to an unfortunate blind spot. Besides, they shut down what premier biblical scholar and theologian Walter Brueggemann calls the "imaginatively playful," maintaining that "more responsible and faithful interpretations take place at the interface between the *normative* and the *imaginatively playful* . . . between the canonical and the imaginative."[10] Brueggemann's fine definition of the term *inspiration* is helpful here, and Ellul would have agreed:

> We mean to say that God's own purpose, will, and presence have been "breathed" through these texts. Such a claim need not result in a literalist notion of "direct dictation" by God's spirit, as though God were whispering in the ear of the human writer; it is clear that the claim of "inspired" is an inchoate way of saying that the entire traditioning process continues and embodies a surplus rendering of reality that discloses all reality in light of the holiness of YHWH.[11]

Although this disclosure occurs in fits and starts, Brueggemann continues, "it is in the end not domesticated by either human imagination or human ideology—we receive a 'revelation' of the hiddenness of the life of the world and of God's life in the world."[12]

A more critical point for our reading of Genesis concerns the overarching question whether biblical texts report "history" in the modern sense of the term in the first place. Here, too, I find myself in agreement with Brueggemann's view: "the literary offer as a vehicle for religious claim does not rise and fall with critical historical reconstruction, for

---

10. Brueggemann, *Introduction*, xii, emphasis original.
11. Brueggemann, *Introduction*, 11.
12. Brueggemann, *Introduction*, 11.

the literature is not the product of events, but a product of imaginary interpretation."[13]

Ellul interprets the text through the filter of God's self-revelation as liberator, so that when God speaks, he puts humans in a situation of *response-ability*. As responders to God's performative word, they will employ their language, culture, and state of knowledge to express what they understand God to be saying. This explains why there is more than one creation account: the biblical authors shared with the people of their time a collection of stories that reflected their beliefs from various angles.

The same pattern pertains to the OT as an account of a journeying people, of many fresh beginnings after falling back into old ways, an account of second and third chances, of regressions and progressions, so that the revelation of God to Israel appeared successively in history and from different vantage points. These aspects do not line up well with the hypothesis of a verbatim inspiration, as if God had given it word for word to his "penmen." Rather, it is more plausible culturally and historically to believe that when it comes to revelation no single text can be exhaustive. This pattern is well attested in the four Gospels of the NT. Ellul sums it up this way: "What we find is human beings who, one after another, heard certain things from God and who transcribed what they thought they heard using what they knew. That is why we have successive revelations that follow and renew one another."[14]

Before we explore Genesis 1 and 2 more closely, it may be helpful to say a few words about the larger body of OT material to which these chapters belong: Genesis 1–11. This portion summarizes the theological history of the world as it was understood in the ancient Near East, Israel's religious and political environment, where it had been used liturgically in foundational myths that narrated primordial events in which "the gods" were the main actors. These poetic narratives had been indisputably appropriated by Israel from other "older, well-developed cultures" as parallel texts from the great cultic centers of the time show.[15] Regarding their importance for the interpretation of the canonical form they have attained today, Brueggemann notes: "That prehistory, while interesting, is not especially important for theological interpretation of the final form

---

13. Brueggemann, *Introduction*, 4.
14. Ellul, *On Freedom, Love and Power*, 4–5.
15. Brueggemann, *Introduction*, 29–30.

of these texts beyond the important awareness that biblical literature did not exist in a cultural vacuum, but in lively engagement with its context."[16]

Names for God

When reading the first two chapters of Genesis, we must consider the difference between the two names of God we encounter: *Elohim* in Genesis 1 and *YHWH* in Genesis 2. This is important as each name carries its own relevance for interpreting the creation story to which it belongs.

Elohim is a plural form used as a singular (like *mathematics*). It refers to the generic and objective name for God and is related to a plurality of divine names used by the peoples in Israel's cultural surroundings.[17] It means God in the abstract, signifying the distant God who keeps to himself. This hidden God is known only through his acts. So, in the first 25 verses of the Bible it is *Elohim* who speaks. *Elohim* was used by the ancient Israelites when they pointed to the divine powers at work in the cosmos, the powers that awed them with their stunning grandeur displayed in the night sky and with the beauty of ceaseless creativity in nature. Modern scientists stand in the same place today when they, like complexity theorist Stuart Kauffman, propose to call the mystery of this awe-inspiring creativity "the Sacred."[18] Its unnerving presence simply unravels the adequacy of human reason. People who experience this awesome presence may discern God in his works, perhaps experiencing a sense of grandeur at the sight of the latest space photos from the International Space Station, which may bring them closer to God. They realize that there is a Creator, there is a God; they may be impelled toward something they cannot discern in nature, and the Bible tells us what they experience is *Elohim*. Thus, the plurality of *Elohim* seems to invest the central affirmation of monotheism that "God is One" with a subtle expansion in meaning, saying that "God is Oneness."

The second creation account no longer speaks of *Elohim* but uses a different name for God, *YHWH*. In the first account, the focus was on knowing who creates and God's relation to the universe as a whole. In the

---

16. Brueggemann, *Introduction*, 30.

17. Ellul, *On Freedom, Love, and Power*, 19. See the section on "Ancient Symbols of Divinity" in chapter 7, where I point to the archaic roots of *El* and their prehistoric Semitic antecedents.

18. Kauffman, *Reinventing the Sacred*, xi.

second account, it is no longer a matter of creating, organizing, directing, but of the relationship between God and humanity. This is where the second account begins and sees everything from that perspective. Far from being anthropocentric, this account asks a different theological question, which demands that from now on everything is to be related to humanity.[19] *YHWH* is the God of the personal relationship, the God of saving grace and intimacy, who, in a certain measure, gives himself over to people, although within limits. God gives his name, *YHWH*, but the tetragrammaton is unpronounceable: it is a name that no human being can utter. God reveals himself but remains hidden at the same time so that we are faced with a curious dialectic: without revelation we are free to imagine anything we like about God and be as precise about it as we please. We can even claim we know this God. But when God does reveal himself, everything we have claimed about God is invalidated.[20]

On the other hand, when God enters a personal relationship with a human being, it is not *Elohim* who speaks, but the one who gives his name. In personal encounter we meet the one who reveals himself. This is *YHWH*, the one who is at the very heart of all life, the one who hands over his name as he did with Abraham, with Moses, and with all the prophets. This God, who establishes contact with the person he chooses, can only be known by revelation. *YHWH* cannot be grasped or "known as the Being of beings unless he gives himself to you. Elohim, on the other hand, can be known through his works."[21] In short, the appearance of *YHWH* in the second creation account invokes the God of the entire OT tradition. In other words, there is, theologically speaking, a considerable distance between the first and the second account.

Composition of the Text

The widely accepted historical theory of how the first five books of the Bible, the Pentateuch, came to be written holds that at least three groups of editors were involved as authors. As noted above, God is called *Elohim* in the first chapter and *YHWH* in the second; in each part the language and style vary. In the texts that feature *YHWH* an older grammar is used, and when combined these texts produce a coherent whole. The same may

---

19. Ellul, *Freedom, Love, and Power*, 42–43.
20. Ellul, *Freedom, Love, and Power*, 42.
21. Ellul, *Freedom, Love, and Power*, 42.

be said for the parts that use *Elohim* exclusively. Then there are sections where *Elohim* and *YHWH* alternate. Here the language and style also vary from part to part. The grammar of these sections differs from the others; among other things, it is more recent. Owing to their rigorous style, these sections are thought to be the work of theologians and taken to belong to the Priestly tradition. The Yahwist tradition is thought to be the oldest, dating from around 900 BCE, and the Priestly tradition from around 700 BCE, with the so-called Elohist tradition between the two, around 800 BCE.[22] The text we have today was the work of later redactors who integrated the the often fragmentary material around 600–500 BCE.

More Issues of Interpretation

"In the Beginning ... "

Genesis 1:1 is usually translated as "In the beginning God created the heavens and the earth." This traditional rendition, so familiar in the English-speaking world, is an unfortunate mistranslation that misses the true meaning of the Hebrew. The mistake was first published with the King James Version in 1611. Judging from the Latin Vulgate—the Latin version of the Bible of the fourth century, traditionally thought to have been the work of Jerome (347–420)—and from the Greek Septuagint of the second century BCE, the wording "In the beginning ... " seriously misrepresents the original. The original Hebrew reads *Be'reasheet*. While it can mean "in the beginning of," it cannot mean "in the beginning." In his book *The Hidden Face of God*, Gerald L. Schroeder comments:

> The difficulty with the preposition "of" is that its object is absent from the sentence; thus the King James Version merely drops it. But the 2100-year-old Jerusalem translation of Genesis into Aramaic takes a different approach, realizing that *Be'reasheet* is a compound word: the prefix *Be'* "with" and *reasheet* a "first wisdom." The Aramaic translation is thus "With wisdom God created the heavens and the earth."[23]

Such an opening places the emphasis of the first creation account on an entirely fresh footing, acknowledging divine wisdom as a fundamental

---

22. Ellul, *Freedom, Love, and Power*, 12–13.

23. Parallels of this idea are found in Pss 33:6; 104:24 (Schroeder, *Hidden Face of God*, 49).

building block of the universe. The creation is expressly declared to be the work of wisdom in the Psalms, in the book of Proverbs, and in Jeremiah, while the book of Job ensures that we do not miss the deep interconnection of wisdom and goodness, for in God's absolute wisdom he realizes the self-expression of his love. As we will see in the remainder of the book, this wisdom is revealed not only in the intricate life processes, including the emergence of consciousness in the cosmos, but supremely in the person and work of Jesus of Nazareth.

As implied above, the first phrase in Genesis 1, *Be'reasheet*, does not describe a point of departure but a setting; it may mean "within the beginning" or "to begin with" or "first of all." Taking other usages in the Bible into account, Ellul believes that it may also be taken as a literary device signaling something like "The narration can now begin."[24]

The God Who Creates

When the authors composed the text we are considering, their principal concern was the revelation of the God who creates, the relationship between the Creator and humanity, and how this relationship came about, as noted in chapter 1. Unquestionably, this focus distinguishes the text from the cosmogonies of Babylon and Egypt. Moreover, it points to a God who is above the fray of mythical cosmic forces that the nations surrounding Israel had venerated as divine. For our purposes, this point prompts two questions: What did the authors of the first account perceive about the God who creates? And to what extent is that perception so incompatible with the dynamic cosmological picture modern science presents that it must be ruled out?

To answer these questions, let me note first that the Hebrew permits a variety of translations pointing to a multiplicity of meanings. To illustrate the difficulties of grasping the meaning of the received text, Ellul quotes an excerpt from a translation offered by Fabre d'Olivet, a nineteenth-century French Hebrew scholar:

> At-the-first-in-principle, he-created, Ælohîm (he caused to be, he brought forth in principle, HE-the-Gods, the-Being-of-beings), the-self-sameness-of-the-heavens,

---

24. Ellul, *Freedom, Love, and Power*, 23. Modern rabbis take *be'reasheet* to say "When God began to create . . . ," leaving room for more to come (cf. see several YouTube presentations on the subject).

and-the-self-sameness-of-earth. And-the-earth was contingent-potentiality in-a-potentiality-of-being: and-the-darkness (a hard-making-power)-was on-the-face of-the-deep (fathomless-contingent-potentiality of being) and-the-breath of-HIM-the-Gods (a light-making-power) was-pregnantly-moving upon-the-face of-the-waters (universal passiveness).[25]

Ellul considers this a good translation, for "it reflects the depths of meanings of the Hebrew words and phrases."[26] For our inquiry, we note what the authors of the text perceived: God as "Being-of-beings" did not create a set of finished products but brought forth a "contingent potentiality of being" that has room and freedom to be itself and even explore creatively its built-in potentiality. Such an insight into the nature of God's creation is far more fluid and thus compatible with a contemporary understanding of the world than a so-called plain reading of the text that has led generations of Christians to accept it as scientifically true because it is taken as God's words on the matter.

Elohim Speaks

If the first two verses of Genesis 1 tell us that it was *Elohim* who created and the *tohu wabohu* (an expression without specific meaning and without linguistic roots) and the *tehom* (a plural made by doubling *tohu*), then the third verse shows us *Elohim*'s first act: *Elohim* speaks.[27] Remarkably, *tehom* is followed by a highly important biblical term, "the waters." This term has a significant meaning in the OT that in the culture of the day was easily understood. "The waters" refer to the powers of annihilation that seek to undo the creation in the same way water dissolves sun-dried mud bricks. Yet, at that stage of the revelation, we are never told outright that *Elohim* created from nothing. This conception had

---

25. d'Olivet (1767–1825), *La Langue Hébraïque Restituée*, 25–26. Astonishingly, this language verges on that of modern physics describing quantum states where reality exists only potentially in superposition prior to the event when potentiality "descends" (falls) into actuality in the collapse of the wavefunction.

26. Ellul, *Freedom, Love, and Power*, 19.

27. According to Ellul, the Jews strongly refused to conceptualize what was before creation, hence the use of *tohu wabohu*, an inexpressible term. *Tehom* has other usages in the Bible that refer to the depths of the sea, "where the big fish live"; although it may be translated in a variety of ways, "abyss" seems to be an appropriate fit (Ellul, *Freedom, Love, Power*, 20–21).

gradually emerged within Judaism, making its first documented appearance around 150 BCE; from there it later entered Christian theology.[28]

The belief that creation happens through divine speech acts is one of the fundamental characteristics of biblical revelation. When we read, "And *Elohim* said . . . ," which recurs throughout the first creation account, we encounter, however, an important ambiguity. God creates by means of a word, by something that is distinct from God and yet is also God himself. The usual verb employed is *dabar*. Being a word and an action at the same time, this word carries a double meaning, and the Jews refused to distinguish between the two. It can, therefore, be translated "God speaks" as well as "God acts." In Jewish thought, words represent a power that exerts a twofold effect: it brings order and establishes a relationship; and when God speaks both occur. We may go as far as to say that when God brings forth something like light or water, he brings forth something other than himself, while imparting something of himself at the same time, just as we do when we speak. Whenever God creates by acts of speech or by his breath, he imparts something of himself. With this act, the dual aspect of divine speech comes into view; it is action and revelation in one, a duality we find running through both the Old and the New Testament. On the one hand, it foreshadows the incarnation; on the other, acts of speech establish a relationship with humanity that invites dialogue by offering freedom to the listener to accept or reject it.[29]

Eretz (Heb. Land, Earth)

Some have argued that the creation accounts appear at the beginning of the Bible because creation ranked first in Jewish thought; the contrary is true. In the Hebrew tradition, the primary concern was not the doctrine of creation but the salvation of humanity and the covenant of God with Israel. An equally misplaced assumption is the idea that the authors intended to instill a belief in creation by providing information about how creation happened (as the first account seems to do). According to Ellul, such thoughts were far from their concern. Their purpose was to teach about the God who revealed himself to Israel and that this God was the

---

28. See the apocryphal book 2 Maccabees 7:28. The Creator's activity consisted not only in creative speech acts, but also in separating and ordering a mass of undifferentiated stuff.

29. Ellul, *Freedom, Love, and Power*, 22.

one on whom they and the entire creation depended. Moreover, when they speak in verses 1 and 10 about the earth, and that it was created, they do not teach a cosmogony. The Hebrew word for earth is *eretz*. When the text was written in the seventh century BCE, for the Jews it meant "the earth of Israel." Because the land God had given them was of the utmost importance for the Jewish people, it is mentioned at the beginning. Further, to distinguish their creation text from those of the Babylonians and Egyptians, the book of Genesis—far from presenting a cosmogony—reveals the God of the land of Israel and of his relationship with its people.[30]

"Here Is the Good"

We read in Genesis 1 how *Elohim*, at several stages of creation, evaluated what he had created. Current translations of the phrase simply proclaim: "And God saw that it was good" (Gen 1:4, 10, 12, 18, 25). Relying on a literal translation from the Hebrew by André Chouraqui, Ellul is convinced that the current English wording is too weak.[31] A better translation would read: "*Elohim* sees: 'O, here is the good.'" This translation indicates that every time God acts something good appears. Also, the text even seems to suggest a hint of wonderment on God's part. Yet "the good" is not in the things themselves, but in what God says. Hence, these exclamations are not the result of comparisons with an abstract good but the affirmation of an original deposit that God placed in a distinct time and place and in his own way. Created entities like the land and the sea are good because of God's declaration. Therefore, every element of the natural order carries a positive connotation. If we took this declaration seriously, it would deny us humans the right to treat the natural order as we please.

This raises an important question about the meaning of the affirmations "here is the good" and "the very good" (Gen 1:31). Do they support the belief held by many Christians that the creation was once perfect but "fell" because of human transgression? To be sure, the notion that God initially created a perfect creation constitutes a fundamental plank in the belief structure of many Christians. But such a position is not supported by the biblical text; the text is quite clear that good is meant and not perfection. What's more, we read in the second creation account that God describes Adam's solitary state (before the creation of the woman) as "not

---

30. Ellul, *Freedom, Love, and Power*, 18.
31. Ellul, *Freedom, Love, and Power*, 26. Cf. Chouraqui, *Entête (La Genèse)*.

good" (Gen 2:18). In other words, it is not inconsistent in God's economy to call some part of creation "not good" while others are good. The sky is good and so is the earth. We cannot draw the conclusion that movements of tectonic plates and their consequences (earthquakes and tsunamis) or weather events like hurricanes are "evil" because of their (from our point of view) catastrophic effects, let alone that such effects were absent from creation prior to the "fall." They were part of God's creation and thus belong to his glory. Like all things, they must be seen in the context of their relationship to one another and to the whole of cosmic existence.[32] After all, God is the creator of all that exists, and all that exists depends on his creativity.

Moreover, Hebrew thought is not dualistic. It does not split the world into a material world and a moral/spiritual one whereby the latter is good and the former is not (as in Greek culture). For the Jews they are one, as are matter and spirit, hence Ellul's view that matter and spirit are linked just as word and action are linked. It is precisely this perspective from which we must read the extraordinary affirmation of the human creation, whose appearance in history God declares to be "very good" (Gen 1:31). It is an affirmation with extraordinary implications for the theological meaning of who we are as human beings.

"Let the Earth Bring Forth..."

The first Genesis account does not show us a Creator who acts like a magician and brings forth predetermined creatures or fills space with predesigned objects the way one fills a container. Rather, we see *Elohim* calling on certain created things, the sea and the earth for instance, to bring forth living creatures. As God invites created entities into the process of creation, he imparts self-creative powers to his creation, a power-sharing process that implicitly involves the entire cosmos. By saying "Let the earth (or the waters) bring forth..." (Gen 1:11, 20), God seems to create room for creaturely realms to fulfill the creation as they bring forth vegetation, mammals, ants, butterflies, fish, as well as microorganisms. This means that God created every creature with a certain autonomy that God will not violate, but also with room for diversity (mutations).

---

32. For a comprehensive exposition of this view, see Southgate, *Groaning of Creation*.

Thereby all creatures bear their own unity of being to form the kind of ecological coexistence we see everywhere in the natural world.

We note that in sharing his creative power with creaturely realms, God seems to limit his omnipotence. By entering such a relationship with the created world, God appears to redress the serious relational imbalance with his creatures, an imbalance that could not be any greater. What does it mean to relate to a God who has the power to create a universe as vast as ours and sustain it over billions of years? How do we relate to a deity who has the ultimate power to annihilate, to reveal and conceal, the power to be present everywhere, and the power to withdraw and hide? How does one relate to such a God? In the first Genesis account, two clues present themselves. God ordains created things (not just humans) to "bring forth," to participate in the creative process, to become stakeholders in the creativity that promotes life. In the divine *letting be*, God invites creaturely realms to be cocreators with him. By letting be, he creates a cosmos that among other things brings forth life on at least one planet in a manner deeply consonant with the Christian experience of God's humble self-emptying love. Letting be creates being and involves letting go, which, in human experience, is a greater power than defending or holding on. This divine attitude of letting be toward the creation is paradigmatically reflected in the NT parable of the Prodigal Son: the father in the parable acquiesces to the son's request to hand over the inheritance prematurely and lets the son go, keeping a vigil of love until his return. We will see the same stance in another place below, when God blesses the creation.

Time

As we saw above, in Judaism—contrary to any other religion in the surrounding culture—God created by means of his word. We also saw that in Jewish thought the creation was an event in time, not a space or container filled with objects. If the surrounding culture began with space, the Jews began with time. The world was happening in time, constantly breaking out in further developments (as reflected in the biblical wisdom literature), so that the Jewish understanding of creation was not at all what others thought it to be.

Therefore, when we read that God appears in the first five days of creation, we see God in action: he gives it form and separates light from darkness, water from land, land from the sea. In each act of separation, he gives shape and form to what is formless, making expressible and comprehensible what is *tohu wabohu*.

A similar point may be made in relation to the Hebrew word for day (*yom*), which can also mean a finite period. Recalling in passing the many fruitless attempts to refute evolution by parsing the meaning of *yom* and the volumes that have been written to quantify a day's duration, I want to stress that none of these arguments is relevant for the meaning of the text. Again, the interest of the Genesis authors was entirely elsewhere. Far more important in Jewish thought is that the day begins with evening followed by morning, presaging a pattern frequently encountered in the Bible whereby the negative element is followed by God's positive act. First, there is night, then comes the light; first there is *tohu wabohu*, then comes life and humanity. This pattern contrasts with the Greek and Roman approach—the model for Western culture—where day begins with sunrise and ends with sunset. We begin with life and end with death. In Jewish eyes, the text deals with a counterintuitive reversal of the cultural model: you begin with death and end with life. It leaves to one side the fruitless squabbles about whether a biblical day was 24 hours or thousands of years long. The egregious misreading of the text, especially when it is used to argue against evolution, could not be more obvious.[33]

While the text speaks of a certain order and periodicity and holds it in high esteem, it does not tell us what "actually happened," as so many authors of the young-Earth-creationist movement have falsely asserted in their attempt to disprove Darwin. It should be clear by now how wrongheaded this enterprise has been from the start.

"According to Their Kind"

In several places, the Genesis accounts state that species of plants and animals are to propagate "according to their kind [species]." This traditional phrasing too is based on a mistranslation of the Hebrew. The latter, Ellul asserts, does not imply a predetermined outcome but a point of departure, so that the phrase should be translated "with view to . . . ," an important nuance implying that everything is to propagate *with a view toward* the creation of its species. In this case, the cosmos and our living world are far more fluid and dynamic than readers of the Genesis

---

33. Ellul, *Freedom, Love, and Power*, 23–24.

texts have traditionally thought possible. What had contributed to this static view was the long-held belief that the creation was complete and that therefore nothing new could be expected in a "finished work" (Gen 2:18). Although the Hebrew text suggests that our universe is a dynamic and open system, I want to reiterate that no convergence of the biblical with the modern scientific description of the world is intended, for the modern description itself is an emergent phenomenon in its own right.[34] What I do maintain, however, is that using Genesis as a weapon to refute the scientific description of the universe by claiming the Bible knows better and alone provides an all-time, valid, step-by-step report of how creation happened amounts to a blatant misreading of its revelational intent.

God's Dialogue with Himself

Next, the creation of humanity is preceded on God's part by a dialogue between God and himself. This inner dialogue seems to suggest that prior to creating humans God summoned the cosmos as a witness to what is about to come forth, at the same time instilling the dialogical principle in humanity that is intended to govern the divine-human relationship at the apex of creation. While this dialogical principle is of the utmost importance in every age, it attains special significance in our time given the man-made ecological crisis we are facing globally.

God Blessed Them

The first creation account ends with God's blessing of the sea creatures, the birds, and all the living things (Gen 1:22–24). By addressing them this way, God sets them apart from the rest of creation, which he had already declared good. The Hebrew word for blessing (*bârak*) in its multiple meanings is highly significant here.

The first meaning is to kneel. This seems strange because we are accustomed to thinking that the one who gets blessed is the one who kneels before the one who bestows the blessing. In blessing all things, the Creator bows himself before his creatures. This presents us with an image of God who in kneeling takes the stance of a servant, upending all power relations that seek to assert superiority of one over the other. This image

---

34. Ellul, *Freedom, Love, and Power*, 25.

reminds us of a passage in the Fourth Gospel, where Jesus kneels before his disciples, washing their feet (John 13:3–14).

The second meaning of *bârak* is to proclaim salvation, which may also be understood as being loved by God. Hence, what God brings to his creatures in this stance of service is not only the affirmation that they are good but also that they are loved.[35] This theological understanding is further enlarged when we consider the human species. On the "sixth day" God blesses the man and the woman, who are created in "the image and likeness of God," so that, as a species, humans correspond more than any other creatures to the Creator. Created as God's counterpart that transcends all differences between men and women, between one ethnic group and another, humans are empowered, as conscious creatures, to choose their posture vis-à-vis the material world of which they are a part. We humans are enabled to assume the very posture of the Creator, who kneels in humble service before the creation. If we did, we would indeed give effect to the image in which we have been created, as servants and stewards of the creation.

In the foregoing I have shown that it was not the intention of the biblical authors to inform us about *how* the creation happened, because their concerns lay elsewhere. I also hinted at the compatibility of the biblical revelation with a dynamic understanding of the creation. In what follows, by looking more closely at God's creative activity in both accounts and focusing on key relational clues, we will see more compelling evidence for why the claim of modern cosmology for an evolutionary universe does not conflict with a biblical understanding of creation. Ignoring for now the huge differences between the two creation accounts as well as the fact that both accounts seem to conflict with current scientific knowledge of the universe, we just want to note that the first account presents God's creative activity known by the subdivision of "seven days" (Gen 1:1—2:1–3) and the other (Gen 2:4–25) sets out, in a different way, to recount the history of the heavens and the earth as created by God.

The First Five Days of Creation

Here we arrive at the traditional account of the creation proper. Ellul stresses the way the Jewish view of creation contrasts with Greek and

---

35. *Salvation,* although an ambiguous term, can be understood as being loved by God and responding to this love.

Babylonian beliefs. To begin, in Jewish thought there is no original act of creation "from which everything else flows and before which nothing was."[36] The act of God concerns relationships. We see God appearing in his creation, separating light from darkness, water from water, land from sea, and so on, at the same time giving form to what was formless, and rendering comprehensible what was inexpressible or *tohu wabohu*. After all, for the Jews the burning issue is not how anything exists, but how the individual human is to relate to it, and how to make sense of it all. The point is, once God calls it into being and names it, the individual can know what it is. This perspective ties back to three planks of Jewish thought: God creates by his Word, there is nothing before that Word, and we must regard creation in this light from then on. It also explains why it never occurred to the Jews to construct a theory of creation. They were not interested in how the cosmos worked; that the stars and other bodies in the sky were lights was sufficient. Unlike other cultures that began their creation story with space, the Jews began with time, a point critical for our inquiry. Ellul writes: "The conception of the world as an event that was happening, evolving, and constantly breaking out is further developed in Proverbs, Ecclesiastes, the Song of Solomon, and so on, with the result that their account of creation is not at all what the Greeks and other people understood by it."[37] In short, the Bible does not present us with a static picture of creation but with a dynamic one that changes with time.

The Creation of Humans

The creation of humans occurs on the sixth day. The text opens in verse 24 with a remarkable flashback on a previous creation event, the creation of animals on the fifth day. This second reference to the creation of animals may be puzzling at first glance. Recalling that in Hebrew repetition means emphasis, the text seems to suggest that the next step in creation will somehow be continuous with the creation of animals (v. 24 → v. 25 → v. 26). Yet, prior to this step there is a pause as God dialogues with himself, deliberating on his intention to create humans in his own image. It is followed in verse 27 by these famous words: "So God created man in his own image, in the image of God he created him, male and female he created them." What is not clear is what to make of these introductory

36. Ellul, *Freedom, Love, and Power*, 23.
37. Ellul, *Freedom, Love, and Power*, 23.

signals preceding the creation of humans. After all, in the lead-up to human creation we find a renewed reference to animals, while the text contains two mentions of the image of God in which humanity is created. It follows that humans were created as ambiguous creatures. On the one hand, humanity is continuous with the animal world, yet, on the other, as the fruit of God's specific self-communication, humanity was to be "in his image and likeness." Humans were thus able to hear what God speaks and dialogue consciously with him as God's counterpart. In short, humans are part of the animal world and yet distinct from it.

In referring to the creation of humans, the text does not command propagation "according to its kind." This, Ellul argues, may be interpreted to signify that human existence is not a matter of speciation in the way the animal world multiplied. Rather, humanity was to be a single type of animal whose uniqueness is derived from the fact that *Elohim* created Adam "in our image." To make this latter term intelligible, much exegetical energy has been spent. Two often-used examples will suffice to show the direction these efforts have taken: humanity is the microcosm of God (who is the macrocosm); humanity is the image and likeness of God because it is endowed with personality, free will, and intelligence. Yet, Ellul argues, these and similar views fail because they neglect what the text makes very plain: that God is man and woman. What comes into view is not an undue emphasis on sexuality but being a plurality, two in one.

Seventh Day and God's Rest

The first creation account concludes on a celebratory note. The seventh day of creation is no longer a workday, but the crown of creation. Indicating completeness and achievement, the number seven is important. Since the text omits any reference to "evening and morning," we are made aware that creation has entered an open-ended period, which, according to Ellul, is the time of cosmic and human history. Structurally, then, in the first six days the cosmos appears, followed by human history after humanity has been created.

Many English translations of the Bible render the Hebrew *shavat* as "rested," but a more accurate translation would be "abstained." Theologically, the text informs us that God "abstained from his creative work" on the seventh day. As Ellul puts it, "God steps back from his creation and no longer acts. It is not that God stops being the creator, but that he no longer makes new things."³⁸ Those who think of God in terms of being an impersonal "first cause" would find this notion disagreeable. But since God is not abstract, in abstaining he acts like a person and not like an impersonal Providence. If he was a "first cause," God would have to go on acting endlessly. We find instead that God, in deciding to rest, manifests his freedom. Since this decision to abstain takes place within history, the text implies that God no longer interferes with creation but grants it freedom. By abstaining from acting in history, God grants freedom to both the cosmos and to humanity to become what they might become. God does not direct everything; indeed, he abstains from acting to the point of resting. Rather, when God created humanity, he created a species through which love is possible, and as soon as this step is completed God steps back to allow his creature the freedom love needs, for love without freedom is unthinkable. This decision to step back does not render God indifferent, for he does not abdicate responsibility; rather, he grants freedom to his entire creation to become itself. And while present, he is present at a certain distance, leaving open the possibility of coming into the fray of human history and working in it, but only through his dialectical relationship with his creation.³⁹

The text also teaches us that God provides us—made in his image—with a model for how to abstain from interfering with creation, and to cease controlling and manipulating the world as we find it, at least once a week. This example is especially relevant in a technological age that prides itself on having created a global system of ceaseless production. Humans are not reduced to objects that must conform to divine fiat but, in their freedom to make their own decisions, are invited to enter voluntarily into the perfect freedom of divine love which is "the rest of God."⁴⁰

All about Meaning

The First Account

If the foregoing demonstrates anything, it is that the revelation of Genesis 1 is all about meaning. This account is not meant to provide information on how creation happened, and even when it offers us events and

---

38. Ellul, *Freedom, Love, and Power*, 36.
39. Ellul, *Freedom, Love, and Power*, 37.
40. Ellul, *Freedom, Love, and Power*, 38.

facts this is only to convey their meaning for the people concerned—that is, first for the people of Israel, then sometimes for the surrounding nations, and later generically for all humanity. The centrality of meaning cannot be overstated, of which events and facts are the outer clothing. This understanding will guard against the error of equating reality with truth. When we assume that God created a reality and then provided us with facts about it in the Bible, we commit a serious error of interpretation. Rather, the events and other data presented in the Bible are there to serve the primary purpose of conveying meaning. As all knowledge and human abilities and energies were the gifts that stood under the sway of a faithful Creator, the Bible knows nothing of so-called worldly matters. All had meaning, and to find it was a matter of reading it properly based on prayerful theological inquiry. Keeping these elements in mind, let me summarize:

- Like all biblical texts, the Genesis account is not divine dictation but an "imaginative remembering" of the prophetic voice in ancient Israel as this people perceived God's voice within their specific historical-cultural context.[41]

- This context differed widely from our modern conception of history.

- YHWH, the God of Israel, was believed to be the creator of all that exists, who had revealed himself through his "performative word" that gave rise to acts of creation.

- God revealed neither a metaphysics, nor a philosophy, nor a system, but something quite different: *God reveals himself in personal relationship with the people of Israel.*

By probing this self-revelation of God more closely as we perceive it from Genesis 1, I have strongly suggested that the meaning of the text does not contradict the new scientific understanding of the cosmos. On the contrary, based on the biblical text, God's work in creation is far more open and internally dynamic than most Christians believe. I believe that many of my readers will find this departure from the "plain" reading liberating, as it allows them to live with greater confidence that in adopting an evolutionary view of the creation they have not become unfaithful to the revelation. One of the strongest pieces of evidence is the richly relational side of the biblical narrative. The text speaks of power sharing,

---

41. Brueggemann, *Introduction*, 1.

of an attitude of *letting be*, of the Creator's servant stance vis-à-vis the creation, of the divine-human relationship marked by dialogue, and of the withdrawal of the Creator from acting in history to make room for human freedom for love's sake.

The next section will provide more evidence for the claim that the biblical text in no way debars a scientific description of God's creation. To the contrary, it reveals the creation's open and dynamic nature, thereby laying bare the senselessness of the century-old evolution-versus-creation controversy.

The Second Account

Commentators have always been aware that the first and second accounts contradict each other in many ways. The first features water from the start, whereas the second begins with dry land. In the first, animals are created first and humans second. In the second, humans are created first and animals second. Adam is instructed to name them and see whether there was any helper for him among them. In the first account the universe is created first and humanity last; in the second humans are created first while their habitat, the garden, is mentioned second. Even on a superficial reading one gains the impression that the first account is presented at a higher intellectual level than the second, while the second shows more detail. Also, textual studies have shown that the second is believed to be much older than the first, yet it appears in second place.

But that is not all. As mentioned, one of the most striking features of the Hebrew text is that the word for God differs in these two accounts, each pointing to a different characteristic. In the first account, the word for God is *Elohim*, a plural form, signifying the generic name of God, the distant deity who keeps to himself and is recognized as the creator of the world. When we are awed by the grandeur of the cosmos, what we encounter in such experiences is *Elohim*, according to the Bible. This is not the God of personal relationships!

In the second account, we meet another word for God, *YHWH*. This is the name of the God who reveals himself in personal relations. When God enters personal relations, he gives his name. He reveals himself as the one who is at the heart of all life and who hands over his name. In *Elohim* we encounter the God who can be known by his works, whereas we cannot know *YHWH* unless he reveals himself by giving himself to us in his name, the unpronounceable tetragrammaton *YHWH*, which has been taken to mean "I am who I am" or "I am here." In the culture of the day, especially in Jewish thought, having a person's name meant having been given access to the spiritual interior of the person, allowing one to act on that person. By giving himself in his name, *YHWH* accepts that he will be acted on by the one to whom he has revealed himself, yet within limits indicated by the unpronounceable nature of the name. In other words, the second account shows us a paradox: the God who reveals himself also hides.[42]

Inquiring into the meaning of this paradox, Ellul observes that if God does not reveal himself, we have only the natural phenomena to go by to understand God's character and are otherwise free to imagine him any way we like. On the other hand, when he does reveal himself, we are faced with the disconcerting fact that everything we previously thought about God is false and that this God is hidden in the most unlikely places.[43] In sum, then, the second account is exclusively about *YHWH*, the God who is both hidden and revealed. One might ask, perhaps in a speculative manner, whether it could have been this attribute that prompted the ancient authors to retain in a single text the older account alongside the younger to ensure that corresponding revelatory elements are preserved.

Since the second account no longer speaks of *Elohim*, but only about *YHWH*, we understand why this text begins with humanity. If we saw in the first account humanity in relation to the cosmos, we see in the second humanity in relation to the relational God, *YHWH*. This important feature of Genesis 2 does not reflect a primitive anthropocentrism but a platform from which to address theological questions about humanity.

There is first the environment in which humanity is placed, the garden of *Eden*, in which there are two trees: the tree of life and the tree of the knowledge of good and evil.[44] The man may eat freely the fruit from all the trees in the garden except from the fruit of the other tree. The issue at stake, as Ellul points out, is not objective knowledge of good and evil since from a biblical perspective there is no reality called "good" and another called "evil." There is simply no good in itself. Good is what God

---

42. Ellul, *Freedom, Love, and Power*, 41–42.

43. Ellul, *Freedom, Love, and Power*, 42.

44. Two aspects of meaning are worth highlighting: the English word "knowledge" is too weak, while "affirmation" would be a better fit; the meaning of *Eden* "designates sexual communion encompassing sensual pleasure and joy" (Ellul, *Freedom, Love, and Powe*, 45).

does and says. From this perspective, the problem arises when humans begin to assert themselves and strive to attain the good apart from God, for apart from or outside of God there is no good. Hence, the dire warning: "The moment you eat the fruit from *this* tree, you will die." This reveals our freedom as well as our problem: we either say what God says, or we say something else. But when we reach for the ability to discern good and evil by itself, we place ourselves outside the relationship with God and begin to displace him and his relationship with our newly acquired knowledge.[45] By thus replacing God with our own judgments about good and evil, we cut ourselves off from the source of all being, hence the warning: "The moment you eat of it, you are dead." The text is commonly translated, "On the day you eat of it, you will die." This reading, however, sounds like an imposed penalty or punishment, but this is not so. God, like a good parent, simply announces the consequence of an act that leads to harm: "When you touch a stove burner, you will be hurt." In the text, God warns that from the moment you aspire to possess the godlike power of determining what is good and what is evil you are already dead. This is not a matter of punishment. Rather, it is the working out of an ineluctable process of severing the link with God by putting us in his place and determining by ourselves what is good and what is evil, as at this point living in freedom has become impossible. What the text reveals is that humanity can only live in freedom when it remains in communion with God and when this communion flows from a relationship of love.

We arrive at an identical point when we consider the issue of service as it is presented in the second account. *YHWH* puts humanity in the garden of Eden "to till it and keep it" (v. 15). Some commentators have questioned the logic of this work mandate as the garden was already fruitful and, as there were no external enemies, there was no point in guarding it. Ellul sees it differently, suggesting that the text is not about real work but about worship (the Hebrew word used here is *avoda*, meaning worship). In other words, humanity was to worship the Creator through their work, simply because God had said so, even if the work was unnecessary in human eyes. For *YHWH* had not launched humans into purposelessness but created them to be his counterparts who respond to him. And since good is what God says it is, humanity was not to make up its own mind whether the garden required cultivation, but simply to do it as an act of worship, thus bringing the whole garden—that is, all of creation—to

---

45. Ellul, *Freedom, Love, and Power*, 45.

worship the Creator. Refusing to work "for nothing" or without human objectives, however, led to a now unavoidable consequence: humanity must work for something that does not arise from a relationship of love and therefore not from freedom. This element is crucial for a proper reading of the second account, for freedom in this sense is not defined as a choice between two equal possibilities. After all, true freedom is not a matter of doing anything we want. It is defined by the relationship of love and only by that relationship, such that, theologically speaking, freedom and love are braided inseparably together like the double helix of the DNA molecule. Love under duress is an oxymoron. The crucial point of the second account then is that Adam (humanity) has been launched into a relationship of love, and when it is maintained Adam (humanity) will do spontaneously what God says, to the pleasure of both.

According to the second account and contrary to the conceptions abroad in the surrounding religious culture, Israel's conception of humanity's creation never envisaged a master-servant relationship between *YHWH* and his people, but a relationship of love where the stakes of existence were life and death. For if Adam (humanity) did not love God, humanity would love something else and thus stop living.[46]

This takes us to the second most crucial relational element in the second account: the relationship between the man and the woman. But first some general remarks about the divine intent for humanity as male and female. As mentioned earlier, using the biblical text to settle the creation-versus-evolution debate is an unfortunate misuse of the text because the Bible speaks about other things. Already in the first account, we noted that the creation of humanity follows a series of what almost look like flashbacks to the creation of animals (e.g., Gen 1:24). This seems to imply that while humans are distinct from animals, they still belong to the same family. Yet the passage about their propagation does not mention diversity; humanity is regarded as a single and unique animal, distinguished by being created in the image or likeness of God.

This reference has been interpreted by different theological traditions differently. For instance, in Catholic theology it means the presence of free will and intelligence in humans; in Calvin's Reformed theology it means the human person. In the previous section, I followed a more theological line of interpretation when I pointed to the significance of "kneeling" derived from the Hebrew word for blessing (*bârak*). As we will

---

46. Ellul, *Freedom, Love, and Power*, 51–53.

see, Ellul's focus differs from all of these interpretations, with far-reaching implications for our understanding of the male-female relationship. In the first account, we read in Genesis 1:27 that "*Elohim* created humanity [âdâm] in his image [or form], in the image [or form] of *Elohim* he created him [or her], male and female he created them." A close reading can only mean that the image of God is of a man *and* a woman, or if one sets aside the issue of sexuality, a being of two in one. Since the word âdâm is singular and the word *Elohim* is plural, the text says that God is several in the unity of one (the plural treated grammatically as singular) and humanity is created as one in two forms separated as male and female, which is the image of God. This raises the question of the relationship between the two, which, in Ellul's reading, can only be love.⁴⁷

If the fundamental relationship between a man and a woman is love (which Jesus later put in these terms: "The two will not be two, but become one"), this love is both sexual or physical love and spiritual love of the entire being—all at the same time, inseparably and indistinguishably—as the Bible does not distinguish between these two aspects of love. This corresponds to what the text has already taught us about the plurality within the unity of *Elohim*. Thus, if the inner unity of *Elohim* is love, then the image of God is love, which in turn predicates the relationship between a man and a woman fundamentally as love, and the two as one.

Also, in Hebrew âdâm is a collective term and means humanity, not a single individual. Therefore, one cannot assert from this text that all of humanity has descended from a single primordial couple. Rather, when God created humanity, he created an ambiguous being within the context of the creation of the animals and yet unique in that humanity bears the image of God with a vision toward love. When we take *this* perspective so deeply embedded in the Hebrew text, then humanity was created within a larger movement tending toward something that converges on love. Such a vision eliminates the whole creation-evolution debate.

One further point calls for clarification regarding the male-female relationship. When the woman appears in the second account, Ellul points to several surprising features. First, in most known languages the word *woman* derives from a different root than the word *man*. Not so in Hebrew. The word for man is *ish*, which means "master." Now Adam names her *isha*. Ellul interprets it to mean that Adam "transfers this mastery to her so that *isha* . . . will now be master." Far from pointing to

---

47. Ellul, *Freedom, Love, and Power*, 30.

matriarchy, Ellul highlights the importance of the suffix in the Hebrew that turns *ish* into *isha* with the meaning "in the direction of the man." In other words, the man alone is not fully human. It is through the woman that the man becomes fully human.[48]

Interestingly, in the text when Adam meets the woman, he responds with a spontaneous and exuberant cry of recognition as he becomes aware of her and, at the same time, of himself: "This is bone of my bone and flesh of my flesh." What it tells us is that thanks to the woman, a man knows himself. What is more, the Hebrew word for knowing is *y*āda, which means "knowing and valuing the other intimately and participatively as in sexual intercourse," and this knowledge is mediated through and by the woman. Such knowledge not only corresponds to God saying about man that "it is not good" for him to be alone, but also to the fact that God analogously invites humanity into the same close intimacy with himself.

In conclusion, let me add two further remarks. Humans, as a he-and-she, can neither be thought of nor understood as an entity independent of the one whose image they are. There cannot be human existence on its own. However, in being the image of God, humanity is necessarily free, for freedom is the prerequisite to love, which in turn implies autonomy. Yet, since humanity is but an image, it is at the same time not autonomous. Here we encounter the fundamental ambiguity of humanity: free and not free, autonomous and not autonomous, two but one, complete as a being yet not whole as denoted by the context of its creation on the sixth day (in Hebrew numerology, the number six is an imperfection but tending toward seven, the number of completeness). Therefore, in the economy of God, humanity represents an indeterminate factor that implies a certain openness in creation. As Ellul puts it, humanity was not saddled with the responsibility for bearing the good but appears to have been sent down the path of an open-ended adventure. At the same time, God addresses humanity—created with the ability to hear the personal words that God speaks (*torah*), always appealing to a coworker in a personal relationship. This sets humanity apart at the summit of creation to respond to, in dialogue with, the Original in whose image they are created from within a relationship of love, which is the inside story of creation. From this perspective, the meaning of the astonishing diversity in the textual material that comprises the OT revelation becomes clear.

---

48. Ellul, *Freedom, Love, and Power*, 54.

We are not dealing with a systematic, progressive revelation but with a God whose pedagogy is adapted to the culture, time, and place, even to the individuals he is addressing as he seeks out humanity in love without overwhelming us.[49]

On this view, let me reiterate that there is no providence in the Bible; in Ellul's mind, the idea of providence is another terrible theological error, for God does not drive history from behind like a chauffeur drives a car. Instead, God comes into the middle of things as perfect love, on his own terms and at his own chosen distance, not to control what is to happen in the world, but "by acting towards us as a person who leaves the field to us."[50]

Conclusion

The key question this book seeks to address is how to live as a thoughtful Christian with intellectual integrity in the age of science without being unfaithful to the biblical revelation. On matters of creation, Christians have traditionally claimed Genesis 1 and 2 as their master text. Since it was understood as the Word of God, these chapters described factually what happened. Suggesting otherwise meant undermining the reliability and authority of the Bible. Trusting scientific evidence for an evolutionary cosmology and for biological change in history was anathema as it represented intolerable compromises with nonscriptural sources. Propagating such views needed to be rigorously disproved with the weapon of a "plain" reading of the text.

Having surveyed the Genesis accounts in some detail based on Jacques Ellul's reading of the Hebrew text, we found that the "plain" reading of the text does not do justice to its meaning, and any claim based on such reading is liable to lead to serious theological errors. In support of this conclusion, let me assemble the most salient points to show one more time that using the biblical text to settle the creation-versus-evolution debate amounts to an unfortunate misuse of the text simply because the Bible speaks about other things.

Like all biblical texts, the Genesis account is an "imaginative remembering" of the prophetic voice in ancient Israel as this people perceived *YHWH*'s voice within their specific historical-cultural context,

---

49. Ellul, *Freedom, Love, and Power*, 109.
50. Ellul, *Freedom, Love, and Power*, 37.

which differs widely from our modern conception of history.[51] After all, the God of Israel revealed neither a metaphysics, nor a philosophy, nor a systematic account of creation, but something quite different: *God reveals himself in personal relationship with the people of Israel.*

As we probed the self-revelation of God more closely, we saw in Genesis 1 that God's work in creation is far more open and internally dynamic than most Christians believe. One of the strongest pieces of evidence is the richly relational character of the biblical narrative itself. It speaks of divine power sharing with creation, of *letting be* as the Creator's attitude toward created realms, of the Creator's own servant stance vis-à-vis the creation, of dialogue with the human creation, and even of withdrawing from action in history out of love to make room for human freedom. This deep relationality is further revealed in the relationship between the man and the woman, presented as the essence of the image of God in the unity of the two as one.

Regarding evolution, in the first creation account we noted flashbacks to the animal world, implying that while humans are distinct from animals, they still belong to that family tree. Also, since âdâm is a collective term for humanity, this text cannot be used as evidence for the assertion that all humanity has descended from a single primordial couple. Rather, in creating humanity, God created an ambiguous being within the wider biological context of the animal world that is nonetheless uniquely distinguished by bearing the image of God with a vision toward love. From this we must conclude that if humanity was created within a larger creative movement tending toward something that converges on love, the whole creation-evolution debate simply collapses. This conclusion is further corroborated by the developmental dimension and openness to which the Hebrew text points at the summit of creation, implying a certain open-endedness in the creation as a whole. This touches the inside story of creation. When God addresses humanity, he is appealing to a coworker (neither a slave nor a puppet) whose freedom God respects and who stands in a personal relationship with him. Humans, as bearers of God's image, are set apart to respond to and dialogue with the Original from within a relationship of love.

But alas, the history of humanity and of Israel testifies to the tremendous historical and cultural turbulences that have accompanied God's project. Yet, in Israel's case, God constantly changed his pedagogy,

---

51. Brueggemann, *Introduction*, 1.

accommodating culture and people. So pervasive is this element that we must regard it as a fundamental feature of the entire revelation. Hence, we can trust a faithful Creator to do the same in our time, showing us his amazing and breathtaking creativity and love, even in the age of science.

❧ 3 ☙

The New Cosmic Story
A Narration

YOU AND I LIVE inside a grand story, the new and epic story of the universe! Over the last 200 years, scientists have revealed with a high degree of certainty that this story began with a ripple smaller than an atom 13.8 billion years ago. In this chapter, we glimpse the marvelous birth of the universe, of galaxies and stars, of the Earth, and of life on Earth. It begins with a narrative like the old creation story. By writing it this way, I became deeply aware of how young and incomplete this new story was. Not only are we discovering more of the universe's complexity and beauty, but, more importantly, the journey of the universe itself is far from finished. So, we must hold the story gently. If we grasp it too tightly, we will harm it as we have harmed the ancient story by believing it was chiseled in stone, fixed and unalterable for all time. If we grasp the new story as hard as we grasped the old, we will surely continue to harm the Earth. Wasn't it our misreading of the old story that caused us to believe everything existed simply for our benefit, for us to use and exploit, without regard for the rest of creation?

When we look closely at the new story, we also behold what many scientists are recognizing: in addition to the visible physical dimension, what they describe possesses an invisible metaphysical one, not least because the cosmic processes that created countless vast structures also brought forth butterflies and bluebells, koalas and kangaroos, and in us humans the capacity for consciousness, which this book presents as the seat of divine revelation.

The Birth of the Universe

When the universe was born, it was a beginning like no other. There was no time, no space, no matter, and there were no eyes to behold the moment when, in an ecstatic trillionth of a second, infinite energy exploded out of nothingness. From a ripple, smaller than an atom, the fabric of space and time flashed forth in seething vibrations as the newborn universe inflated faster than the speed of light to a size larger than a galaxy in almost no time at all. This was a detonation like none other, perfect in all its measurements: a trillionth of a second faster, and the universe would have spun out into nothingness; a trillionth of a second slower, and it would have collapsed back on itself. Inflation continued but no longer accelerated; gravity appeared as a fundamental force.

Ultrarapid expansion meant cooling, and the infant universe cooled to near absolute zero. In the process, other fundamental forces became distinct. As they separated, they gave out energy so that the universe, now a millionth of a second old, heated up again and created a cauldron of quarks and antiquarks that annihilated each other. From this seething chaos, only a billionth of the previously available mass survived as protons and neutrons.

After three minutes, from the surviving protons and neutrons, nuclei of hydrogen and helium were formed, yet the formation of *atoms* of these primordial gases took more time, in fact 300,000 years, as the positive protons captured the negatively charged electrons.

As matter and radiation separated, more energy was released in the form of radiation, and the entire universe ignited as a white-hot fireball whose light is still detectable today and is known as cosmic microwave background radiation. At this point, the infant universe was still completely ruled by radiation. The only substance that existed was a submicroscopic precipitate, suspended in the white-hot fireball whose "light" consisted of powerful X-rays and gamma rays far too fierce to permit atoms to form.

In the first few hundred thousand years of its existence, the radiation era of the baby universe, all was uniform, symmetrical, and without much structure. As the universe expanded, and its expansion became a touch slower, electrically charged particles could capture others of the opposite charge to form neutral atoms as the structure of matter.

Gradually, radiation's dominance was replaced by the dominance of matter, perhaps representing the greatest change of all time. This moment of symmetry breaking occurred when the universe was around 500,000 years old. Now all the ingredients of creation were in place—space-time, energy, and the fundamental forces and particles that, over the next 13 billion years, rearranged themselves in "endless forms most beautiful."[1] They transformed what was given into structures of increasing complexity, from galaxies, to stars, to planetary systems, to biotic life, and, eventually, to sentient creatures such as us, the creators of cultural systems that have altered nearly everything on our planet. A new era had begun—the era of abounding and still-ongoing creativity.

At first, for millions of years the expanding universe consisted of an undifferentiated sea of atoms and energy. Then, precipitously, gravity drew the atoms into trillions of spinning whirlpools, vortices of gigantic clouds spinning, eventually as huge protogalactic disks, with even vaster distances between them. Within each cloud, by the same processes of gravitational ordering and shaping, smaller clouds formed that condensed into spheres of enormous mass. Density increased under gravity's pressure, forcing atom against atom so that the interior of these massive gaseous spheres grew hotter and hotter until they ignited. Accordingly, when the universe was about 200 million years old, it gave birth to the first stars.

The first galaxies began to form when the universe was about a billion years old, multiplying enormously over the next 1.5 billion years, so that this period is known as a boomtime in chemical element making. But it would take another two billion years before our mother galaxy would form, the Milky Way galaxy. Did the universe know what it was doing? Was there perhaps a mysterious inkling of how exceedingly fecund its grandiose unfolding would turn out to be?

The Travail of Galaxies

If asked, "How many stars are in the Milky Way galaxy?," astronomers will say that this is one of the most difficult questions to answer, for not

---

1. Darwin, *Origin of Species* (1859), end of last paragraph quoted in Carroll, *Endless Forms Most Beautiful*, 281–82.

all stars shine brightly enough to count from Earth. Since even a sun-size star would be too difficult to detect, astronomers use other methods to estimate the number of stars in the galaxy. According to the best estimate, it contains 100 billion stars plus the remains of many that have expired since their formation. While we relish the bright band of starlight in the night sky, from which our galaxy derives its name, flashing forth beauty is not a galaxy's primary task; rather, it is to crash stars. When gravity first formed our galaxy in the Local Group of galaxies that resides on the outskirts of the Virgo cluster, our solar system did not yet exist. The chemical elements making up the sun and its planets today show an average age of about 10 billion years. They were produced in now burned-out stars that scattered their residue into space, from where gravity collected them to be recast in new forms. These recycled ashes of burned-out stars eventually furnished the material substances that compose the "dust of the Earth," the oceans, and all living creatures and their biological evolution, including the development of human bodies and minds.

The great gas and dust cloud of the galaxy flattened and moved disklike around its center, now understood to be a supermassive black hole. The protogalactic disk took the shape of a whirlpool with flaring, spiral arms. Exploding stars sent gargantuan shockwaves through the galaxy, in turn causing gas and dust clouds to collapse and to ignite to form new stars. When the spiral arms move through such a zone, a chain reaction of star explosions is unleashed, and the production of chemical elements in the galaxy skyrockets. The galactic region where our solar system would appear (around 4.5 billion years ago) experienced this kind of shock treatment for billions of years. Every time a massive star exploded, the region benefited from the residue left behind.

As turbulent as these events may be at the scale of the stars, the galaxy itself is moving in relative tranquility, unlike some other assemblies. One factor aiding this undisturbed condition is the universe's expansion, which makes intergalactic space less crowded and collisions of galaxies less likely. Our nearest galactic neighbor is the Andromeda galaxy, and astronomers predict a collision with the Milky Way in about four billion years. Overall, the stellar productivity of all galaxies has been phenomenal. Today, their living offspring are estimated to number 70 billion trillion (7×10^{22}) stars and counting—as the universe gives birth, every day, to around 275 million new stars.

The Significance of Stars

Stars are born everywhere in the universe under the pressure of gravity from huge gas clouds that float within galactic formations. These enormous gas clouds that weigh tens of thousands of times more than our sun could be dubbed stellar nurseries, for that's what they are, as they give birth to thousands of stars all at once. If we take the process of star formation in slow motion, this is what we see. Gravitational pressure first generates slightly denser portions of the cloud that then fall back on themselves, further increasing their density. As gravity's unrelenting grip keeps squeezing these huge spherical masses of hydrogen atoms, their collapse escalates faster and faster. Under such runaway conditions the core of the sphere ignites and begins to glow—a protostar is born. Over the next several hundred thousand years, the surrounding gas merges more and more with the baby star so that its glow becomes gradually more exposed until it becomes visible. Some stars draw the surrounding gas and cosmic dust into protoplanetary disks that then encircle the new star. Over time, planetary systems form in a like manner: vortices are squeezed into spheres that condense into planets by gravity's inexorable pressure.

After about 10 million years, the new star reaches adolescence. It is now able to generate its own heat in its core through nuclear fusion that converts the hydrogen of its interior into helium.[2] The adult life of a star (like that of our sun) lasts a long time, in the order of 10 billion years. During its entire life cycle, a star not only generates heat and light, but through nucleosynthesis it converts hydrogen into all the other chemical elements we know. In short, most of the visible universe, including the Earth, is made of elements once formed in the cores of stars that since have died. Yes, stars die eventually when they run out of hydrogen fuel. As its mass declines, the crush of gravity eases, and a star bloats into a red giant. All stars, irrespective of their mass, go through this stage, although eventually the way a star dies depends on its mass. Most stars die quietly, slowly disembodying themselves as they puff off the chemical elements they have produced like smoke rings into space. Other stars die more explosively in immense eruptions or supernovas. These, by all accounts, are the most powerful energy events in the universe after the big bang. A

---

2. In a medium star like our sun, the temperature in the core is about 13.6 million degrees C. Thermal radiation and convection currents transfer heat to the star's surface, where the temperature is about 6,000 degrees.

supernova will reach the brilliance of a billion suns for weeks, outshining entire galaxies, flashing forth in one short energy burst more energy than our sun in its entire lifetime.[3]

Even on their deathbed, some stars can produce elements heavier than iron, such as uranium and gold, and finally, in their last gasp, bequeath to the universe their legacy, the very stuff of life: carbon, silicon, and oxygen; sodium, calcium, and phosphorus; copper, zinc, iodine, and iron; and, indeed, all the natural elements of the periodic table.[4] Cosmologists Brian Swimme and Mary Evelyn Tucker write: "It is remarkable to realize that over immense spans of time stellar dust became planets. In the earliest time of the universe this stellar dust did not even exist because the elements had not yet been formed by the stars. Yet hidden in this cosmic dust was the immense potentiality for bringing forth mountains and rivers, oyster shells and blue butterflies."[5]

One of these second-generation stars is our sun. Over four billion years ago bands of matter orbited in its protoplanetary disk. Some of it solidified into small spheres, the planets, nine in all, one of which is our Earth. Back then, the expanding universe had reached about half its present size.

The Birth of a Planet

When the Earth was very young, asteroids, comets, and meteors bombarded it for long periods, adding to its substance. At one stage, a huge asteroid ripped a piece from its still-unstable body and the moon was born. It now circles the Earth, stabilizing its orbit at a seasonally variable distance from the sun. Within the next billion years, Mercury, Venus, Mars, and Pluto turned into solid rocks, while Jupiter, Saturn, Uranus, and Neptune have remained spheres of elemental gases without much change in four billion years.

---

3. A supernova may take a week to reach maximum luminosity and may stay rather bright for several months afterward.

4. The helium gas in popular party balloons was created when the universe was three minutes old, but the air we breathe contains oxygen, which originated in a rather complex two-stage nuclear process. This process first fused three helium atoms to produce one carbon atom, which fused with another helium atom, yielding an oxygen atom.

5. Swimme and Tucker, *Journey of the Universe*, 36.

How different the Earth is! Its size enabled gravity and the electromagnetic force to work in tandem, creating a magnetic field that shields the Earth from fierce life-destroying solar radiation. At the same time, its position in relation to the sun gave it a temperature range in which complex molecules could form in its oceans, setting the conditions for the emergence of an entirely new form of creation, biotic life—albeit a billion years later.[6] How life formed on the planet is, in itself, a story of epic proportions that will concern us toward the end of the chapter.

Light from the Past

In 1964, two scientists, Arno Penzias and Robert Wilson, had developed a supersensitive antenna to investigate faint radio signals from outer space. Because their experiments required unprecedented levels of precision, they had eliminated all possible interferences like the effects of radar and radio broadcasting. They even cooled the receiver to near absolute zero to prevent it from generating unwanted noise. When they examined their data, they discovered a persistent and steady yet mysterious hum in the microwave range, 100 times more intense than they had expected. This noise came evenly from across the sky. Checking their equipment again, even removing some pigeons and their droppings from the antenna, Penzias and Wilson found the noise persisted. Concluding that it was coming from outside our galaxy, they could not explain the source.

At the same time, researchers from Princeton University were working only 60 kilometers away, searching for microwave radiation. One of them, Robert Dicke, had predicted in 1960 that cosmic microwave radiation existed and should be detectable. His coworker, Jim Peebles, was preparing a paper on the possibility of finding the leftover radiation from the initial explosion that gave the universe its beginning. Hearing about it through a friend, Penzias and Wilson suddenly realized the significance of their observations and, together with Dicke and his coworkers, concluded that what they had discovered was indeed the cosmic microwave background radiation (CMBR), the leftovers of the initial cosmic explosion. Penzias and Wilson were awarded the Nobel Prize in Physics in 1978 for the discovery that substantiated the big bang theory. In other words, their discovery of a form of "light" that originated around 14 billion years

---

6. Tyrrell, "Oldest Fossils Ever Found," https://news.wisc.edu/oldest-fossils-found-show-life-began-before-3-5-billion-years-ago/.

ago provided scientific proof that the universe had a beginning, that it is structurally more or less the same in all directions (isotropic) and expanding, and that its size has been increasing ever since.

Why is this knowledge important? The answer is straightforward: because we ourselves are deeply rooted in cosmic history. Indeed, we carry the history of the universe in our DNA, for every elementary particle that constitutes our molecules has a multibillion-year history. But I am getting ahead of myself. Before we consider the cosmic breakthrough that brought forth living forms from the ashes of stars, we must first immerse ourselves in the chaos that accompanied the birth of planet Earth.

Cosmic Breakthrough—Living Forms

If we asked a group of biologists today what divides living and nonliving systems, we would be hard put to find two definitions that would come close to resembling each other. The difficulty of defining what life is lies in the diversity of its manifestations, in the complexity of the underlying biochemical processes, and, last but not least, in our fundamental ignorance of its genesis. We simply don't know how life began. Modern physics has even raised the question of whether the common division between living and nonliving matter is a false dichotomy. Be that as it may, the origin of life is a blank spot in the pages of cosmic history. Like the big bang, it is a mystery. Yet we know for sure that life is there, and we also know something about the timing of its emergence on Earth. So, let's try to look a little more closely at the momentous breakthrough event in cosmic history, when living forms emerged on Earth from the stuff of the universe or from chemical elements spewed into intergalactic space by dying stars in supernova explosions.

As the Earth slowly gathered more elements from the solar disk, it added more substance to its thinly forming surface and to its toxic atmosphere of methane, hydrogen, ammonia, and carbon dioxide. These gases roiled in violent turbulence above, while seething magma churned beneath the crust. Like a huge cauldron, the young Earth boiled and cooled over millions of years. Giant volcanoes erupted everywhere, spewing molten lava as well as massive clouds of dust and water vapor into the atmosphere, while meteors and comets constantly bombarded the Earth's surface. Electrical storms of immense ferocity were unleashed that lasted millions of years and drove the planet through new chemical opportunities toward radically new phases of emergence. These hellish conditions would have vaporized any accumulation of surface water as they blasted the tiniest beginnings of an atmosphere into space. On the usual time scale, we find ourselves now approximately 10 billion years after the big bang. The Earth was lifeless and void; there were no continents, only large volcanic islands like those of Hawaii or Iceland. Yet amazingly—and we don't know how—only 200 million years later biotic life seems to have gained a foothold on the planet in the form of microbial life, possibly first near hot deep-sea vents.

As the Earth's surface cooled enough for liquids to remain, our planet was ready for new possibilities. Its crust, floating on the seething magma beneath, experienced ceaseless rainstorms that filled the seas, separating the volcanic islands. New possibilities arose when churned-up materials from below, created in faraway stellar furnaces and remolded deep within the Earth's boiling interior, were thrust up through volcanic eruptions onto the surface and through deep-sea vents within the oceans, allowing new and complex molecules, in endless combination, to appear. Latent mechanisms of self-organization that permeate every region of the universe encountered molecular structures and free energy ready to make use of them, and so set the stage for a momentous event in planetary history: the formation of the first living cells 3.8 billion years ago. Tiny bacteria emerged, establishing a beachhead on the planet and seeding the Earth with living things. Strikingly, some of their kind have persisted, in an unbroken line, to the present.

As one would expect, life's first protocellular structures were the most fragile and delicate creatures in this epic story, yet most essential for advancing it. Their descendants, early prokaryotic cells, multiplied and in time filled the seas, feeding on the energy-rich chemistry of the hot, young Earth. They relied on the continuation of these chemical creations for their ongoing food supply. However, as the Earth settled after a few million years into chemically less turbulent conditions, the supply slowed down. When consumption outstripped supply, early life was threatened. Just in time, a mutation arose to avert disaster. New forms of single-celled microorganisms appeared. These organisms were able to feed on the wastes of dead or dying cells, and even on their body parts, advancing life further, for as wastes and decaying bodies became food for living cells they spun the first threads of the web of life among living creatures: the food chain and the possibility of an ecology had emerged.

Then, another remarkable mutation occurred. Cells "invented" photosynthesis; through their own molecular structure, they learned how to convert the energy particles (photons) of sunlight into food. How they did it is astounding and points to the deep interconnectedness that pervades the material world. Microorganisms on Earth interacted creatively with the energy coming from the central star of the solar system 150 million kilometers away (which converts 600 million tons of hydrogen and 600 million tons of helium every second into heat and radiation energy).

But development beyond single-celled organisms would be unthinkable had it not been for another amazing event in the history of the planet. Geoscientists call it the Great Oxidation Event. About 2.4 billion years ago, the Earth and its shallow oceans experienced a near-catastrophic rise in atmospheric oxygen. Until then, the Earth's atmosphere was devoid of oxygen as most of its atmosphere had been formed by the process of outgassing. Gases trapped in the Earth's interior were released during extensive volcanic activity, spewing vast quantities of toxic gases and water vapor into the atmosphere. As the young planet cooled further, atmospheric temperatures also dropped, causing a global deluge that formed the oceans. As a result, carbon dioxide levels also dropped, so that the atmosphere was now dominated by nitrogen, remaining without oxygen for more than another billion years. Without what happened next, most likely none of us would be here; no dinosaurs would have roamed the Earth; no bird or bat would soar above; no shark fin would ply the oceans; no mammal would suckle their young. Microorganisms alone would exist because without oxygen, for all we know, the breathtaking potential for life's creative emergence would have come to a dead end.

How this great event happened is not yet fully understood. However, there is wide agreement among scientists that cyanobacteria that had been around for a billion years were largely responsible for generating free oxygen (the prefix *cyano-* comes from a Greek word meaning dark blue). To explain what happened, we must backtrack and delve deeper into how cells learned to draw energy from the sun. The early cells had begun to twist the porphyrin ring (a macromolecule of organic compounds, part of their inner structures) into a net of other organic materials. It allowed them to capture individual photons from sunlight as it traveled at the speed of light through the surrounding waters and convert their energy into food. By using sunshine, water, and carbon dioxide, they were able to produce carbohydrates plus free oxygen as a byproduct. Today, more than two billion years later, all plants on Earth employ cyanobacteria (or chloroplasts) to do their photosynthesis for them. There is also much evidence that oxygen had become a significant component of Earth's atmosphere around 2.4 billion years ago. The ratio of sulfur isotopes changed in rocks dating from that time, while iron oxidation occurred on the seafloor, leaving red iron bands as markers indicating that seawater oxygen had drastically increased. The new atmosphere brought an additional advantage: protecting organisms from the sun's harmful ultraviolet rays.

But not all organisms could tolerate rising oxygen levels. Prokaryotes (cells without a nucleus) perished en masse, giving way to a new form of life: cells called eukaryotes, or cells with a nucleus. This nub of the cell's core protected vital genetic information and contained tiny organelles with specific functions, one of which was to process oxygen. Eukaryotes thrived in this new environment.

Yet mysteries remain. We don't know, for instance, what caused oxygen levels to rise high enough to enable the evolution of animals, or how atmospheric oxygen reached present levels of around 21 percent and remained stable for two billion years, making life as we know it possible. But let's return to the big picture.

To understand the processes that led to life, it is essential to understand two closely related concepts—emergence and self-organization. The concept of emergence has become central to an understanding of the behavior of complex systems in the modern scientific context. We speak of emergence when the properties of a system cannot be deduced from its constituent parts alone. For instance, we cannot explain the beautiful and symmetrical patterns in snowflakes from the properties of water molecules. Likewise, we cannot explain the functions of multicellular organisms from the properties of the individual cells that constitute them or predict the behavior of a flock of birds or school of fish from the behavior of individual specimens. Thus, emergence has been defined as "the arising of novel and coherent structures, patterns and properties during the process of self-organization in complex systems."[7] To understand these dynamics, we must understand how systems cope with change. We ourselves as living systems live as part of the system of the universe that evolves and grows. At first, science was mainly concerned with systems in equilibrium. That changed in the 1970s and 1980s, when Russian physical chemist and Nobel laureate Ilya Prigogine showed that open systems

---

7. Goldstein, "Emergence."

do not strive for equilibrium, but have the possibility of importing free energy, which keeps them off balance and fosters change and growth.

These systems, by engaging in an active exchange with their world, use what they find for their own renewal. To regulate the behavior, these systems operate with two feedback loops, one negative, the other positive. These two loops use information differently. The negative loop monitors departures from stability to render the system stable; the positive loop amplifies responses to extremes until the system cannot cope and begins to shriek, calling for adaptation and growth. In other words, systems can overcome their demise (entropy) by exchanging energy with the environment. Prigogine's work on the evolution of dynamic systems has shown that *disequilibrium* rather than equilibrium is necessary for systems' growth. Systems that possess the internal properties needed to reconfigure themselves go by the name of self-renewing or self-organizing systems. The point is their resilience rather than their stability.

This self-renewing dynamic leads to an increase in the system's organization and occurs in biological systems from the subcellular to the level of ecosystems, in physical processes and in chemical reactions. Astrophysicists observe the same dynamics of self-organization in star and galaxy formation. For instance, the typical spiral-arm configuration in galaxies occurs only when a galactic cluster reaches a critical size of approximately 100 million stars. Self-organization is well known in chemistry and molecular biology. In the former, it is called autocatalytic; when observed in the latter, molecular self-assembly occurs where molecules adopt an arrangement on their own, without outside direction. It operates, for instance, in the self-assembly of lipids to form cell membranes, in the assembly of the double helical structure of DNA through hydrogen bonding of the individual strands, and in the assembly of proteins from multiple folded protein molecules (so-called quaternary structures).

There are characteristics common to all these examples: radical novelty (properties that were previously not observed in the system in which the new wholes maintain themselves over some time), coherence (new properties appear as the result of a dynamic process), and evolution of properties (when they can be perceived and said to be ostensive). None of these properties can be reduced to underlying laws of physics, for the laws themselves do not generate anything; they merely describe the regularities and relationships that occur consistently in nature. Because of its importance, let me demonstrate the meaning of *emergent* by referring to the exceedingly remarkable properties of water. While all of these properties are essential for life, not one can be reduced to, or predicted from, the gaseous components of water, hydrogen and oxygen (2 H + 1 O = H$_2$O).

These two gases do not, on their own, exhibit the same properties as they do when they are bonded together to form water molecules. An oxygen atom on its own consists of a nucleus and four bonding pairs of electrons. It is thus electron negative (electron negativity measures an atom's ability to attract a bonding pair of electrons). Therefore, when an oxygen atom bonds with two hydrogen atoms, only two of its four electron pairs are satisfied. This leaves the oxygen atom with two unbonded electron pairs. Although hydrogen is electrically neutral, it becomes slightly positive after bonding with oxygen, which polarizes the water molecule: it is now positive at the hydrogen end and negative at the unbonded oxygen end. This gives water molecules the ability to bond with other water molecules and to stick to surfaces other than water.

In other words, the electrical polarity of the water molecule results in some new, and very useful, properties: (1) *cohesion*, whereby water molecules are attracted to each other to form droplets that fill rivers and oceans; (2) *adhesion*, whereby water molecules are attracted to other surfaces, making water an excellent solvent capable of dissolving and transporting more substances than any other compound, thus sustaining metabolic processes; (3) *surface tension*, whereby water molecules pull toward each other at the interface with other substances (e.g., air), with the result that the smallest possible surface of water is exposed; and (4) *capillary action* emerging from both cohesion and adhesion as water molecules stick to each other and to other surfaces, and so follow each other, even climbing to considerable heights against the gradient of gravity. In living organisms, right down to the level of cells, water transports nutrients and wastes of metabolic processes in solution and, by processes of diffusion and osmosis, water ions can penetrate tissue and cell membranes, which is vitally important for biotic life. Moreover, water is the only substance on the planet that exists in all three states: liquid, solid, and gas. In the liquid state, water molecules continuously bond with and release one another. In the vapor state (steam), molecules move too quickly to allow mutual bonding, while in the solid state (ice) the molecules are tightly bound into rigid configurations that allow for only minimal movement between them.

Living Cells—A Wonderland

Stepping inside the outer membrane of a living cell is a jaw-dropping adventure. As unassuming as it looks from the outside, its interior is packed with a breathtaking array of interactive organic-chemical machinery in perpetual motion undergoing 500 quadrillion chemical reactions per second. A writhing ball of microtubules, one of three inner filaments that make up the dynamic inner skeleton of each of our 100 trillion cells, is in constant motion. Microtubules also function as intracellular highways carrying a cargo of proteins to where they are needed. Every cell produces 2,000 proteins every second, continuously every day, and every protein string is made up of several hundred amino acids, a molecule of some 20 atoms.

The inner work organization of each cell is equally amazing—selecting, directing, organizing, checking, joining, and transporting 500,000 amino acids consisting of 10 million atoms folded into preselected strings and other shapes, and shipping these custom-made proteins to where they are needed inside and outside the cell body. In addition, the entrance to the cell through its membrane is carefully guarded by an army of sentinels at countless portals to keep unwanted intruders out, but letting wanted products in. They lock and unlock the portals when molecules with the right key appear. Signals to open the cell membrane can come from within the cell when molecules are needed as building material, or from outside when neighboring cells send the right kind of prompt.

This tireless and highly complex intracellular activity and its seemingly intelligent coordination raises a critical question. What enables strings of unfeeling and unthinking atoms of carbon, hydrogen, oxygen, nitrogen, sulfur, and phosphorus—the fundamental components of biochemistry—to give rise to such precisely orchestrated behavior? After all, the laws of physics and chemistry that govern atomic interaction are unalterably fixed. But what we see in the interior of cells looks like a vigorous, needs-based, information-rich, even purposeful enterprise. Then there is the storage of information associated with the genetic code in the DNA molecule or the structure of the 20-nanometer-thick (a nanometer is a billionth of a meter) cell membrane. It is made of lipids that form a highly flexible skin that retains the memory of its original shape: when punched, it snaps back, and when punctured, it reseals, all while going on with its usual work without a pause. Or take the cells of our heart muscle.

They have an extra-large number of receptors that allow the stimulus hormone adrenalin to pass through to increase the energy output of the heart when the body senses danger. So charged, the heart rate rises dramatically and the heart pumps more oxygenated blood into the vascular system, powering the body into action.

So much more could be said about the wonder of inner cell activity and its organization, from cell division to the work of neuron cells and more. While some of this activity may arise spontaneously in nature—for instance, lipids can form naturally—it is difficult to overlook the lavish information density that needs to be reached to jump from naturally occurring lipids (hydrophobic organic compounds) to living cells. For one, the blueprint for proteins necessary for the construction of cells does not occur naturally. Their origin remains a mysterious paradox: cell functions depend on proteins, but proteins depend on functional cells, as they are products of cell metabolism. Yet, not long after the appearance of water on the planet, both came into existence at the same time.

Then there is, in the cell's interior, the most complex multilane, multistory transportation system of literally thousands of major highway intersections and crossovers stacked on top of each other, all in a 30-nanometer sphere. This jungle of roadways consists of microtubules, ultrathin fibers that also serve as the inner skeleton that gives the cell its flexible shape. What travels on these microhighways are newly made molecules stacked in pouch-like blisters covered in motor proteins, from the point of production inside the cell to where they are needed. Motor proteins, energized by the fine-tuned release of glucose, do the work of chemical hands that grab and pull the load along their assigned path every second of every day by the thousands. This breathtaking complexity and finely balanced, high-precision biochemical activity within a space 100,000 times smaller than the smallest visible speck is truly awe-inspiring and speaks of a hidden yet humble wisdom beyond our comprehension. This amazing biochemistry has nurtured the symphony of life for billions of years. The human body is comprised of 100 trillion cells each undergoing 500 quadrillion chemical reactions every second of every day. Only recently have scientists begun to understand some of them, and the more they do the more they are enthralled by the wonders they discover of the fundamental mystery of the living world.

Embedded Wisdom?

What I have just presented is a relatively crude description of only a minuscule part of the whole. I have not mentioned cellular waste products and their transportation, the metabolism of glucose and of enzymes, cell division and the work of DNA and RNA, not to speak of the mystery of how the cell can tell one molecule from another. Yet as limited as my description has been, it does draw attention to the underlying complexity necessary for life: the symphonic dance so masterfully performed by the biochemical orchestra resident in each of the billions of cells in our bodies, indeed in the cells of all living things, all working toward life's realization, even sentient life, from an invisible and internalized score. Additionally, when life appears, it cannot stop expanding and pushing beyond the status quo. This thrust is usually referred to as the drive for self-transcendence, manifesting in all processes a built-in intricacy that seems to pervade all existence. Do we dare call it "wisdom," even "a miracle"? With this question, I draw attention to an amazing insight that some physicists have felt compelled to express when they have looked at the data without prejudice: the need is disappearing to understand the material realm as distinct or even in competition with the spiritual dimension, for the two are complementary. For instance, MIT-trained scientist Gerald Schroeder believes from scientific evidence that "the universe is the physical expression of the metaphysical," while the German physicist, and successor of Werner Heisenberg as director of the Max Planck Institute, Hans-Peter Dürr understands the material realm as "congealed spirit."[8]

From statements like these we begin to grasp that the cosmos is imbued not only with an all-pervading complexity and all-encompassing unity but also with hugely complex information encoded everywhere in nature, which we can study in the biochemistry of life. The latter manifests itself in the laboratories of science—especially in quantum physics—first as energy, then as matter carrying (nonphysical) information, which points to an underlying spiritual unity. In other words, there is an inside story to the whole, and understanding the puzzle of existence and the wonders of the material world in terms of *this* story more fully requires tools other than the tools of science. Reducing everything to material-empirical explanations will simply not suffice, for in addition to scientific insight we need spiritual insight. In a word, science and theology—or

---

8. Schroeder, *Hidden Face of God*, 1–3; notes taken by the author from Hans-Peter Dürr's lecture at the Max Planck Institute, Munich, 1995.

cosmos and revelation, as in our book title—are merely two sides of the same totality.

Conclusion

We began this chapter by affirming that when the universe was born there was no time and no space, neither inside nor outside, neither above nor below. What came forth was the entire universe and it happened everywhere at once. Everything that existed in cosmic history—from helium to oxygen, from bluebells to butterflies, from trilobites to dinosaurs, from *Homo habilis* to modern humans—was included and belonged to that first primordial and unimaginably powerful burst of energy. In that exuberant and ecstatic moment of creation, all the essential parameters were present; some have since dropped off like the shells of booster rockets, while others continue on their cosmic trajectory, urged forward as spacetime expands, as galaxies move away from each other, some disappearing from view as they slip over the cosmic horizon. Yet into this drama life appeared as an emergent property on the surface of at least one planet. In a further stage of emergence, it brought forth human brain tissue capable of begetting consciousness, which by reflecting the built-in creativity and wisdom embedded in the natural order transcends its material-biological dimension.

❦ 4 ❧

Understanding Our Place in the Universe

FROM THE DATA OF modern science, a new and unified picture of the cosmos is emerging, including ourselves as the sentient outcome of biological change over time. Evidence for change is everywhere, from gas clouds to galaxies, from stars to planets and to life itself. Today the sciences are weaving an intricate pattern of understanding that constitutes the prodigious fabric of knowledge across the natural sciences. This has made us increasingly aware of the rich relationality, as well as the internal structure and order, of the cosmos of which we are part.

 Having narrated the birth and unfolding of the cosmos in the previous chapter, I wish to shift the emphasis in this chapter and explore the place of humanity in this new cosmic picture. The starry canopy of the heavens has exerted a deep fascination for us humans from prehistory until now, as if, in a primal way, we have always understood ourselves as cosmic creatures. As true as this may be, we are today confronted by a peculiar tension. During the last 100 years, not only has modern science upended all previously held cosmic conceptions, but technological advances have made it possible for the new cosmic picture to be beamed by satellite into every living room and cell phone on the planet. In other words, for the first time in history millions of people have become aware of their cosmic situation by watching moon walks, Mars landings, and Jupiter flybys on live television.

At the same time, there are scientists who declare that humanity is little more than an insignificant cosmic accident, destined—with all our accomplishments—to disappear from cosmic history. Still, it is tempting to believe that we stand in a special relationship to the universe despite the undeniable possibility of our insignificance.

The purpose of this chapter is to show that the latter view—emphasizing our special role in the universe—is credible, even though critics have dismissed it as a self-protective illusion conjured to shield us against a reality too distressing to bear. Indeed, I move in the opposite direction and draw attention to the significance of our place in the new story. Beginning with a scientific perspective, I also address some theological implications, although their fuller development will have to wait until chapters 8 and 9.

Rediscovering a Cosmic Perspective

Since most of us are unfamiliar with the big picture of the universe, I suggest that we begin by relinquishing the very common idea that the universe has a geographic center; given its enormous size, such considerations are negligible. If we must speak of a center at all, we can safely refer to the point "where the observer is," meaning that we *are* the center when you and I look out into space, reinforcing our innate anthropocentric leaning. As a next step, I suggest that we hand ourselves over to the childlike wonder and enthusiasm so simply expressed by famous cosmologist Carl Sagan when he wrote:

> I went to the librarian (at age five) and asked for a book about stars.... And the answer was stunning. It was that the Sun was a star but really close. The stars were suns, but so far away they were just little points of light .... The scale of the universe suddenly opened up to me. It was a kind of religious experience. There was a magnificence to it, a grandeur, a scale which has never left me. Never ever left me.[1]

Millennia earlier, our ancestors too were captivated by the same grandeur and magnificence and, like Sagan, beheld with quasi-religious fascination the wonders of the night sky. They too developed models to make sense of this awesome world, and to recapture ancient perspectives we will briefly consider some of them. Then, by delving into the amazing

---

1. https://en.wikipedia.org/wiki/Carl_Sagan.

degree of fine-tuning scientists have observed in the precision of the mathematical constants that govern the coherence and intelligibility of our world, we will heighten our own cosmic awareness. Among these observations, we note a most remarkable feature that is easily missed unless it is pointed out to us—the astonishing place of the human-size scale among all the other sizes in the universe, from the smallest to the largest. In this context, I point to some milestones of cosmic evolution, noting chiefly the rise of human consciousness, which yields an astonishing perspective that too highlights the importance of our place in the new story.

To prepare us for this wide-ranging exploration, I first offer a few words on the theory of knowledge, or epistemology, that has governed scientific thought in Western culture.

Theory of Knowledge—A Very Brief Introduction

One of the presuppositions of this book is that a coherent reality undergirds human existence. In other words, the universe—of which we are part—is not self-contradictory. As reasonable as this sounds, this assumption of cosmic coherence (as a universal basis for our understanding) presents us with several logical implications we may not have considered before. For instance, cosmic coherence would imply (assuming identical criteria for truth finding) that understanding gained in one discipline complements the knowledge gained in another. Furthermore, the same principle would oblige us to not just remain open to knowledge from disciplines other than our own but to embrace and integrate it so as to form a more comprehensive view of reality and not reject new knowledge when it clashes with our preconceived ideas. This becomes especially relevant when we address fundamental questions of the Christian faith in the age of science. For example, "What does the existence of self-reflective human consciousness mean for our understanding of the universe?" Or "Where does our persistent drive for meaning come from?" In short, the principle of universal coherence is a key pillar in the quest to integrate the knowledge of science with the knowledge of God— that is, with theological meaning. One of the other presuppositions of this book is that the discoveries of science can become a rich basis for deepening our theological understanding and hence for our faith. After all, theology is the art of bringing the data of faith into dialogue with an evolving world.

From the beginning, the Christian community had to wrestle with the concepts of Greek philosophy. Later, in the Middle Ages, Thomas Aquinas engaged with the philosophy of Aristotle. When the work of Nicolaus Copernicus upset the prevailing cosmology 200 years later, theologians had to struggle again with newly evolving knowledge. The point is that the natural sciences emerged from within the Judeo-Christian worldview, which had taken shape under the influence of Plato's and Aristotle's philosophy. This combination brought forth over time what is known as the scientific method. Its rise is largely attributable to three axioms. Scientists believed that:

- The data they were gathering by observation and experimentation were not meaningless.
- To ensure that the meaning was not arbitrary, certain definitions were needed demarcating how the data were to be understood.
- Truth finding needed a method of verification because natural reality as measured might differ from initial assumptions.

Since Christianity had demythologized the cosmos, the planets and other cosmic objects were no longer mythical or divine. This led to the doctrine of the rationality of God and of nature; it is reflected in the first axiom, which made meaningful measurements possible. The second axiom rests on a proposition that meaning can be defined. This element of Greek philosophy combined with the Christian notion that natural reality is potentially meaningful led in due course to a plethora of meaningful concepts across many levels of scientific inquiry. The third axiom is grounded in the Judeo-Christian judgment that the world is inherently contingent or not necessary, depending on God, who created it freely, not out of necessity. In such a world, the testing of our hypotheses becomes essential. These relatively simple statements undergird the theory of knowledge of Western culture, determining its method, scope, and validity; it circumscribes all knowledge gathering and vouchsafes its coherence and reasonableness.

Because today through the media people are aware of developments in science more than at any other time in history, the spread of a scientifically informed worldview is inevitable. What is sadly lacking in our time is a meaningful theological response. As Heidi Ann Russell has pointed out, "If theology and pastoral ministry fail to engage with these developments, we run the risk of becoming irrelevant in the contemporary world."² This point is of special relevance as mainstream theology has neglected an engagement with the natural world for the last 500 years. Leaving the cosmos out of their thinking since the Reformation, however, had major consequences: it prevented a serious Christian commitment toward a threatened Earth. Today, American evangelical Christians are among the most avid climate-change deniers, epitomized by John McArthur, one of their prominent leaders, who recently preached with much rhetorical flourish that God intended the Earth to be a disposable planet for our benefit.³ Having said that, let us return to the big picture.

Ancient Cosmologies

One of my goals in this book is to identify paths along which the thoughtful believer can relate the Christian worldview to the scientific model of the world. In this context it may be helpful to go back in history and see how people, particularly in the ancient Near East, constructed their cosmologies, including the predominant cosmology of the OT. How the first Christians built their model will be considered in the next section.

The word *cosmos* is of Greek origin and was first used to refer to an impeccably and beautifully made object, artifact, or machine whose elements are so ordered that everything works together in harmony. *Cosmos* is synonymous with order and beauty; the opposite is chaos and disorder. In Greek thought, this word later signified that the totality of existence is an ordered whole. Analogously, the principle that ordered the universe was understood to be essential in giving shape to human lives.

Admittedly, we know very little about the religions of cultures that existed before the first millennium BCE. What we do know, however, suggests that even ancient cosmologies sought to bring all that existed into a single whole. Ancient myths of origin thus told stories about where people came from and about their relationship to the whole. Our ancestors often narrated what they saw and experienced in poetic form and built it into the mythical framework of their culture. Whatever evidence they had from observation and experience was presented from within this matrix of meaning that gave coherence to their lives. In other words, ancient cosmologies reflected what people were familiar with as well as a growing perception of a greater whole behind their tangible world.

2. Russell, *Quantum Shift*, 79.
3. See Introduction, note 6.

Thus, many ancient cultures lived with a relatively simple cosmological picture: a flat Earth with the dome of the sky above into which the celestial bodies and their courses were set. Although their founding stories differ considerably, the civilizations of Mesopotamia and the Levant (Sumer, Babylon, Canaan, and Judea) shared the same cosmological picture. This dome is the "firmament" of the biblical creation story. God had created the firmament by separating the primeval waters; it is pictured as an opening, like an air bubble, within the primeval waters, resulting in the waters above and the waters below.[4]

While our forebears sought to give a coherent account of how humans fitted into the grand scheme of things, these narratives were not mere stories. These early cosmologies—like all cosmologies—served as explanations of why things are the way they are. They also point to the presence of an irrepressible human hunger for knowledge and understanding as well as to a penchant for the transcendent. To underline this latter aspect as a very ancient feature of human culture, let me take a brief detour.

In 1995, the first human-built holy place was discovered at Göbekli Tepe, a hilltop structure near Urfa in eastern Turkey. It spreads over 22 acres and consists of 20 enclosures with paved stone floors. There are 200 limestone pillars each 5.5 meters tall. The remarkable feature of the Göbekli Tepe site is that the pillars are T-shaped; they display carved arms and hands, a belt with buckle at the midsection, and a loincloth below the belt; a necklace decorates the column near the top. Each of these pillars weighs 15 tons. Archaeologists believe that they represent anthropomorphic beings (the T-section may represent the head). In addition to the pillars, several crudely sculpted human heads were excavated as well as a totem stone pillar. There is no sign that this structure was built for domestic purposes. Judging from the finds of decorated stone bowls, pestles, and goblets (preliminary analysis of vessel residue suggests that the vessels contained wine), public feasting seems to have taken place there. This remarkable structure was obviously built for huge gatherings and cultic purposes during the Paleolithic, yet the claim that Göbekli Tepe was a

---

4. Then God said, "Let there be a firmament in the midst of the waters, and let it divide the waters from the waters." And God called the firmament heaven. Then God said, "Let the waters under the heavens be gathered in one place, and let the dry land appear; and it was so" (Gen 1:6–9).

place of "organized religion" that predates agriculture, and may even have given rise to it,[5] has been criticized by more recent studies.[6]

But back to ancient cosmologies. In the well-known Egyptian representation of the firmament, the goddess Nut is pictured as a nude female on all fours arching her star-dotted body above the Earth, providing the canopy often identified with the starry belt of the Milky Way, which can be seen with great clarity on a cloudless night above Egypt. According to Egyptian mythology, Nut swallows the sun in the evening and brings it forth again each morning as the sun travels through her arching body.

For the ancients, the fear of the heavens falling down and bringing back the primordial chaos was very real and led to several mythical solutions. In Egyptian mythology the god Shu holds heaven and Earth apart; in Greek mythology the primordial titan Atlas has this role; while in the biblical story of the Deluge the primordial chaos motif finds a highly dramatic expression in terms of God's judgment and rescue.

In these ancient cosmologies, the canopy was understood as a shelter in which people take refuge against the dark powers of the surrounding chaos. What becomes apparent is the intimate link between cosmology, existential angst, and, in a rudimentary way, a theology of hope. What stands out, however, is that for our ancestors the elements of their cosmology are divine.

In Egypt, the primordial waters and the stars are gods. The god Re rules over the gods and over humans. When humans rebelled, he sent gods to destroy them but then took pity on them; he refrained from executing them but left them with a fractured relationship with the heavens. Then he appointed an earthly king, Pharaoh, to rule as viceroy and as a steward of the divine order. Pharaoh was to do what he could to reflect the principle of divine order, *ma'at*, lest the world descend into chaos. Earth was the center of the universe around which everything revolved, although this realm was not valued, as only the divine realm above was considered to have truth value. Humans only had value in relation to the divine hierarchical order.

The cosmology of the OT was similar. The flat Earth was supported by pillars and surrounded by water, and the dome of the firmament rested on the rim of the disk. Beneath the Earth were the waters of the abyss and *sheol*, the underworld, the place of the dead. The lower heaven was

---

5. Torrey, *Evolving Brains*, 134–37.

6. Dendrinos, "Dating Gobekli Tepe," https://www.researchgate.net/publication/317433791_Dating_Gobekli_Tepe.

the realm of the birds, but also of the planets and stars, while the upper heaven was the abode of divine beings.

While cosmogonies (ancient stories about the origin of the world) were openly mythologically framed, often detailing clashes between warring deities or the deliberations of divine councils and the lives of the gods, we note a new departure in the Hebrew tradition. Despite some similarities with other ancient stories of origins, the Judeo-Christian tradition takes as primary sources for its cosmological picture the two creation accounts of Genesis (1:1 to 2:3 and 2:4–25). Knowledge of creation is further supplemented by material from the Psalms (8, 19, 29, 65, 104, 139), Proverbs (3:8), Job (38–39), and Isaiah (40–55). The great difference between the OT cosmology and other origin stories is that the Hebrews never deified the forces of nature.

As we saw in chapter 2, the focus of the Genesis accounts is on the God of creation, who created a world in which it is possible for humans to live. This is a world that came into existence not through a battle between hostile metaphysical forces, but through the speech acts of a benevolent Creator. While the narrative still speaks in the idiom of the contemporary culture, what is decisive, even revolutionary, is that a new frame of reference has appeared: the creation is given theological meaning. In terms of our inquiry, revelation takes place *within* a culture and does not go beyond what people can handle at the time, for revelation is also to serve the needs of that culture, though this makes it difficult for other cultures to fully grasp the significance and meaning as it was intended in the original context. At the same time, the ancient Hebrews adopted the then-contemporary cosmological picture of the world uncritically. After all, outside their view of *YHWH* and his relationship with them, there was not much for them to call into question. This minimalist view had the advantage of being more readily adaptable to the shift that followed in the second century BCE under the influence of the philosophy of Plato and the cosmology of Ptolemy than it would have been to the more complex Egyptian world picture.

The Cosmology of Classical Antiquity

The cosmology of classical antiquity was characterized by an astonishing diversity of models. What interests us here is less their content and diversity than the underlying metaphysical concepts. From the classical period onward, these entailed three: deity, humans, and the cosmos itself. While these entities were variously conceived in terms of their origin and their relation to each other, the details need not concern us here. In Greek thought the cosmos was central. It was perceived as a perfectly ordered whole and constituted the source of human social and cultural self-understanding, so it deserved the deepest respect. In that schema, cosmology, anthropology, and theology were complementary disciplines. For instance, in the *Timaeus*, Plato (c. 427–c. 347 BCE) admires the cosmic order and beauty created by rational agency for beneficent purposes. Plato's cosmological arrangement placed the Earth at the center of the universe. It was surrounded by seven spheres—the realm of the planets—above which was set an eighth sphere for the fixed stars. The celestial bodies were imperishable and moved consistently in circular orbits; they were regarded as animate and guided by reason. But everything below the lunar orbit (i.e., the Earth) was transitory and subject to change.[7]

In classical antiquity the cosmos itself was pregnant with religious values because the regular movements of the celestial bodies testified to a superior intelligence behind this observable order. Even morality was grounded in the cosmic order. For Aristotle (384–322 BCE), as in Plato's cosmology, the planets (moon, sun, Mercury, Venus, Mars, Jupiter, and Saturn) moved around the Earth in perfectly circular concentric orbits. These spheres were hierarchically ordered. Those farthest out possessed higher dignity and perfection than those nearer to the Earth (proximity to the Earth introduced impurity).[8]

This world was created by a demiurge, a manifestation of the perfectly good and beautiful divine mind, according to Plato. It is even referred to as "God," although Plato never thought of it in theistic terms—that is, the way Israel conceived of the Creator. Because the demiurge, who resided beyond the temporal world, was in every respect the reflection of divine perfection and reality, the world was created in accordance with that image as the best possible world. This also ensured that mind, body, and cosmos were deeply interconnected on philosophical grounds.

Plato also taught that the cosmos itself, as a living entity, is ensouled, and that planets too are animate, even godlike beings. On this basis, humans are charged with the task of using their divinely given reason to bring their own chaotic selves back into harmony with the cosmos. This

---

7. Wildiers, *Theologian and His Universe*, 21.
8. Wildiers, *Theologian and His Universe*, 23.

is a matter of allowing the revolutions of the heavens to guide our faculties when we experience inner states of disharmony. Through learning and imitating the imperturbable and unvarying revolutions of the divine absolute, human variability might be stabilized. The human body, however, is seen as the prison of the soul, which belongs to a higher realm. Human bodily existence can and should be transcended by means of the intellect and by studying natural philosophy, which educates the soul so that it can escape earthly existence and be reunited with the realm of ideas, from which it came.[9] This belief in the ascent of the soul to God was widespread in early Christianity. It drew heavily on Platonic cosmology, in which everything was full of life. While such thoughts originated in a Platonic emphasis on cosmic plenitude and abundance, they dovetailed with Judeo-Christian theism and can still be traced in Christian theologies until the late Middle Ages.[10] What helped was that Platonic cosmology was close to the Judeo-Christian notion of creation. On the other hand, Christian theologians struggled with Plato's belief that matter existed eternally.

We take up the question of early Christian responses to the cosmology again in the next section. Here I want to add a word on their attitude toward critical thought and philosophy. Two prolific authors and church fathers stand out: Tertullian (c. 155–c. 240 CE) and his contemporary Clement of Alexandria (c. 150–c. 215 CE). The former, known for his often-extreme rhetoric, spoke out against a "mottled Christianity" based on philosophy, though he was not really opposed to philosophy and critical thought. In fact, he argued that theology should be based on truth, and theological inquiry was to be done with proofs and evidence and with arguments whose premises were shared by all parties.[11] In his view, there was no room for irrational thought in theology; philosophy was therefore not to be despised by Christians. From Clement's standpoint, just as God had given the law to Israel as a schoolmaster, so God had given philosophy to the Greeks to bring the Greek mind to Christ. Philosophy was not only necessary for an ordered way of life, but because it was an expression of humanity's rational faculty it was unavoidable; as such, philosophy should be done well.[12] For Augustine, the use of philosophy was

---

9. Plato, *Timaeus*; Zeyl and Sattler, "Plato's *Timaeus*."

10. Basil, *Hexaemeron* 5.7; Augustine, *City of God* 11.22–23; Aquinas, *Summa Theologiae* 1.47.1.

11. Tertullian, *Apology* 25.

12. Clement, *Stromata* 1.5; Augustine, *Christian Doctrine* 2.40.60.

a form of spoiling the Egyptians. While the early Christians were critical of some philosophical presuppositions that clashed with their faith, their general disposition toward philosophy was that it played a useful role in the search for understanding.

The Ptolemaic Cosmos

One of the great mathematicians, astronomers, and geographers in the early Christian period was *Claudius Ptolemaeus* or Claudius Ptolemy (c. 100–c. 170 CE). He lived (and died) in Alexandria, the capital of the Roman province of Egypt at the time. His Latin name suggests that he might have been a Roman citizen. He used Greek philosophers and Babylonian astronomical observations, especially Babylonian lunar theory, to establish the longest-lasting cosmological model in history. His only surviving work, the *Almagest,* is a comprehensive astronomical treatise. It presents mathematical techniques developed by Babylonian astronomers to calculate astronomical phenomena and geometric models based on observations. His tabular presentations made it easy to calculate future planetary positions. It also contains a catalog of 1,052 stars.

Ptolemy's complex geometric model of orbits was consistent with the Platonic philosophy of ideal forms. Typically for his time, objects in the sky were projected into celestial spheres based on Aristotle's universe. It consisted of 55 crystalline celestial spheres beyond which the sphere of the Prime Mover existed. The Earth was central and stationary; the planets and the sun moved in circular epicycles around the Earth. Throughout the Middle Ages, the *Almagest* was regarded as *the* authoritative text on astronomy. That it rested on Aristotle's philosophy gave it special authority, particularly in the church, which had elevated an Earth-centered model to the status of dogma.

If judged by its longevity, Ptolemy's universe was the most successful model of all time. Little wonder that the transition to a heliocentric view of the universe turned out to be so arduous and costly. With the advent of empirical science since Galileo, together with the development of better instrumentation and more refined mathematics, Aristotle's epistemology and his theories about substances crumbled and a massive paradigm shift became inevitable. In sixteenth-century Europe the winds of change were blowing everywhere: in religion, the arts, literature, philosophy, and science.

What is important for this book is that Ptolemy's concentric, static, and hierarchical universe formed the background for early Christian thought, and to this day the churches still tend to mold doctrines to this static cosmos as if the shift to experimental science had never occurred, the Earth was still central, and God were seated on a throne somewhere above the clouds; as if Maxwell and Einstein had never lived, and as if the dynamic and expanding features of the universe had never been confirmed. Besides, Christian theology seems unmoved by the fact that the cosmic worldview of Western civilization is in the grip of revolutionary change. This shift in consciousness is the focus of the remainder of the chapter, leaving reflections on theological implications for later.

Newton's Mechanistic Universe

Revolutions tend to overthrow old regimes.[13] This was surely the case when the work of Nicolaus Copernicus unseated the Earth- and human-centered Ptolemaic worldview. While the church opposed Copernicus's heliocentric system, it did not forbid the study of nature. As already intimated, the roots of the natural sciences grew from the soil of Christianity.

Scholastics who understood Aristotle's philosophy of nature as a two-way path—generalizing from observations to universal laws and from there to prediction of particulars—played a major role in advancing the scientific method. One of its key proponents was Roger Bacon (c. 1214/1220–1292), who had studied in Oxford and Paris. He was an admirer of the English stateman and philosopher, later bishop of Lincoln, Robert Grosseteste (c. 1175–1253), and was also influenced by the Arab mathematician and philosopher Ibn al-Haytham (c. 965–c. 1040). Bacon advanced the mathematical analysis of light and proposed the use of instruments for scientific experiments. Philosophically, he was a forerunner of critical realism who rejected Plato's conviction that pure universal ideas and forms existed in another world. Bacon clearly separated knowledge gathered by the scientific method from moral argument and religious belief. He was a keen advocate of a methodical approach to science involving detailed descriptions of experiments, reliance on mathematics, and the use of logic to test arguments.

---

13. This section and the next owe much to the admirable work of Ilia Delio, especially her *Christ in Evolution* and *Unbearable Wholeness*, 5–12, 17–35.

The advance of the natural sciences during the Renaissance encouraged human beings to do their own observations and come to their own judgments free from the shackles of religious authority. With the collapse of Ptolemaic cosmology following the Copernican revolution and a growing consciousness of a heliocentric universe in which the Earth was no longer central, scientific discoveries created a new picture of the world. For the first time in history, it was felt that what was once attributed to God's role in the universe could now be explained by science. The universe itself began to reflect more and more the mechanisms discovered in nature rather than the hierarchy of being that characterized the older cosmologies. Decentering planet Earth also dislodged humans from their central position, with the result that the experience of the world as God's creation was harder to come by.

Into this state of unsettledness stepped philosopher René Descartes (1596–1650), who, although an advocate of the mechanistic view of the universe, also sought to reconcile this new and emerging picture with a belief in God. While God, in the medieval synthesis, was the source of unity, Descartes, in searching for a reliable new center, overcame rising doubt about the centrality of the human person by formulating his famous "I think, therefore I am." With this decisive turn to the self-thinking subject the world was radically changed. With one stroke, every person was now on their own, saddled with the task of making sense of the world based on rational thought alone. Truth ceased to exist for its own sake. As now understood, it stood in the service of subjectivity. What is true for me may not be true for you. What's more, as Ken Wilber remarked, the rise of the self-thinking subject became a substitute for the transcendent self of the ultimate Whole, a move that not only split spirit and matter but also set humans on an internally conflicted path. We cannot help but seek transcendence, but we will now seek it in ways that do not satisfy, in a way that even prevents us from finding it: "Food, money, fame, knowledge, power—all are ultimately substitute gratifications, simple substitutes for release into Wholeness."[14] In short, the new heliocentric world picture combined with Descartes's dualism profoundly disconnected human beings from God and from the cosmos, a constellation for which the mechanistic universe of deist Isaac Newton (1642–1727) provided the cosmological rendition.

---

14. Wilber, *Up from Eden*, 16.

According to Newton's *Philosophiae Naturalis Principia Mathematica* (*Mathematical Principles of Natural Philosophy*), published in 1687, the universe functions like a clockwork, a view that decisively influenced the intellectual world of his time. Based on his science of mechanics, the universe too allowed mathematical predictions as it was governed, like all mechanical systems, by physical laws that operated with mathematical precision. In the seventeenth century, science also began to influence views of creation. Newton believed that God "form'd Matter in solid, massy, hard impenetrable, moveable [and indivisible] Particles."[15] He also theorized that God had created a machine-like universe that operated on mathematical and mechanical principles. Time and space were absolutes; matter was passive and subject to physical laws, for which his laws on gravitation provided a comprehensive and mathematically coherent explanation capable of describing the behavior of bodies with mass. Newton thus provided the tools for a mechanistic explanation of the cosmos by physical laws of force and matter.

Certainly, the widespread acceptance of this mechanistic universe led to an unprecedented increase in observational activity—the describing, measuring, quantifying, and cataloging of astronomical objects—over the next 100 years. This was a good thing. We discovered, for instance, that Newton's cosmos is governed by a great deal of predictability, that the physical world, at least in its large structures, is stable and lawful. For example, we can count on Newton's laws to hold when we want to build a bridge, calculate the gravitational pull of the moon, or send a spacecraft to photograph the rings of Saturn. Newton's laws work brilliantly, but only within limits, as Einstein showed.

As the science of astronomy advanced, aided by improved telescopes, theologians began to debate the theological implications. Newton's God was the great engineer who constructed the cosmic clockwork, and, since space was understood as a container filled with objects, this kind of thinking also began to infiltrate theology, so that God was imagined as another kind of object among other objects that existed remotely "out there." Although God still stood behind the laws of physics, once the universe was launched God could retreat beyond the heavens, for he had become superfluous in creation and in history. A formal, absentee-landlord image of God increasingly took hold that was far from the living

---

15. Cited in Lovell, *Immensities*, 116.

God worshipped by Jews and Christians, whose immanence was closer to us than our breath.

This inert, mechanical world was hostile to life and an ill-suited medium for communication between us and the Creator, stripped of all symbolic gravity that had once bestowed meaning on medieval existence. Now its silent expanse signified nothing. In a word, nature had become irrelevant, and we are still reaping what this mechanistic perspective has sown. In a technological society with a free-market economy, it is the dominant paradigm. Both thrive on the denigration of nature, a disdain for other people, the maximization of wealth, and the highly problematic assumption of unlimited growth in a finite world.

As we will see toward the end of the chapter, the new cosmic story no longer supports the idea that nature exists solely to satisfy human needs and caprice. It invites us to look at the world differently, recognizing that it is intrinsically valuable because the cosmos has existed for billions of years without us and has always been valued by the Creator. In the next section, we will see that the potential for such a change in perspective has been inherent in scientific discoveries going back a century—discoveries revealing a universe whose inner workings turn out to be very different from Newton's static and mechanical scheme. Scientists have discovered that there is a *process* in nature, an unfolding at all levels of the cosmos from the very large to the very small, and together these processes brought forth the complexity of biological life.

The New Cosmology—Toward Fundamental Wholeness

Newton's idea of time and space as absolutes had reigned supreme in scientific cosmology for 300 years. Space was a container where the laws of physics applied equally in all directions. Time too was a constant, flowing unimpeded and at a fixed rate everywhere in the cosmos.

This neat scheme came crashing down almost instantaneously when Albert Einstein (1879–1955) published his theory of special relativity in 1905. He posited that space and time form a single dimension relative to each other. Space and time could "shrink or expand depending on the relative motion of the observer who measured them."[16] According to American physicist Adam Frank, "The Universe was a *hyperspace*, a

---

16. Frank, *Constant Fire*, 146.

world with an extra dimension. . . . In relativity every object becomes four-dimensional as it extends through time."[17]

Further, Einstein's work led to a new understanding of gravity. For Newton, gravity was a force that drew objects of mass toward each other. In Einstein's framework, however, gravity has the effect of structuring space-time, acting as a curvature formed by matter. It not only shrinks or expands the three spatial dimensions depending on their direction relative to the gravitational field, but also shrinks or inflates the flow of time. In other words, space-time is curved in the vicinity of a massive body. This groundbreaking shift in physics led to a new cosmology. Gravity and space-time combine to determine the large-scale nature of the universe and lead to the idea that the universe is evolving such that the properties at smaller scales are engendered at the cosmic level. In short, the metanarrative in the new cosmic story is the cosmic process itself, as evolution at the cosmic level sets the pattern for all other forms of evolution—geophysical, biological, human, and sociocultural.

An astonishing feature of this vast and complex cosmos is its underlying wholeness. What alerts us to this possibility is the absence of solid, measurable building blocks of matter when we look deeper into its structure, say at the quantum level. Here we encounter instead movements, vibrations, and relationships. One of the most eminent physicists, David Bohm, takes this discovery to another level. Classical physics, he maintains, had bequeathed to us the perception that reality consists of parts that make up the whole, while quantum theory teaches us the opposite: that the notion of separately existing entities is an illusion. Bohm sees in the endless conflicts and confusion and in the urgent crises that confront us today the fruit of our failing attempts to live by this paradigm of fragmentation. In contrast, he starts with what we observe at the quantum level, an undivided whole. This unbroken whole constitutes the inward order of all existence, which he calls the "implicate order."[18] What we observe in science is the natural or "explicate order" that is derived from the whole and is a subset of the whole. From the perspective of the implicate order, we no longer look at reality in terms of external interactions between things, but in terms of the internal or enfolded relationships among them. Just as the musical score of a Beethoven symphony is merely an abstraction of the music it symbolizes, every instance of nonliving

---

17. Frank, *Constant Fire*, 147.

18. Bohm, *Implicate Order*.

or living matter is merely an abstraction of the whole while it remains at root a feature of it. At the quantum level, all parts are connected to every other part, yet every part retains its inner, enfolded relationality. Thus, the whole is the basic reality, not the parts. As human beings we seem separate, but at the level of our chemistry—the structures of cells and DNA—we are part of the biochemical world and are deeply connected to all other nonliving and living "children" of the same cosmic process.

Another primary aspect of Bohm's implicate order is movement. He calls it "holomovement," a model of reality derived from the idea of a hologram, a three-dimensional image of an object formed by interference patterns of reflected light beams. In such a holographic image every fraction of the image contains the full information on the whole object. Thus, the term "holomovement" helps us to grasp this fundamental feature of the implicate order, provided we rid ourselves of another illusion grounded in classical physics—that movement occurs step by step along a timeline. For Bohm, movement does not connote one instance after another, whereby the first instance no longer exists when the next instance occurs. Rather, holomovement implies that each instance bears in itself the memory of all previous instances. At the same time, movement means that we can never fully exhaust the properties and qualities of the whole. They are always beyond us, so that "holomovement" also means "endless depth."[19]

Far from being mechanical and static, the new cosmological story turns on a universe that is deeply interactive, relational, and expanding. Additionally, it has a past and, scientifically speaking, an unfathomable future. Today, our understanding of the universe is based on the hot big bang model, according to which the universe is 13.8 billion years old and still in the early stages of its "unrolling" (the original meaning of "evolution"). When astronomers speak today of the universe, they mean the observable universe, this spherical region that comprises all matter visible from Earth or from Earth-launched observatories. The observable universe, with a radius of about 46 billion light-years, is estimated to contain two trillion galaxies. It has been calculated that in 100 billion to one trillion years the cluster of galaxies known as the Local Group, which includes the Milky Way and Andromeda galaxy, will merge into one large galaxy. In that time, if the present model holds and dark energy continues to drive the expansion of the universe, many galaxies will have

---

19. Delio, *Unbearable Wholeness*, 29; also, Bohm, *Implicate Order*, 196–99.

moved beyond the cosmological horizon. Remaining stars will increase their luminosity as they age, but as stars burn out the total luminosity of the universe will decrease in 800 billion years. Yet hundreds of billions of years before this occurs the sun will have burned out and collapsed into a neutron star. This, however, is a distant prospect, for the sun has fuel for another five billion years. As its core diminishes, the central star in our solar system will inflate to become a so-called red giant. In that state it may burn for one billion years. As it inflates, the distance to Earth will shrink, changing the habitable zone; oceans on Earth will evaporate and its surface rocks will become too hot for life.

So much for the new cosmology. Developments in other disciplines have also accelerated in the direction of understanding the world as dynamic processes. In the mid-nineteenth century, Charles Darwin proposed that animal and plant species developed from such biological processes as natural selection, suggesting that living phenomena emerged not as one-off creations but from processes. Darwin did not use the term "evolution," a well-known expression in the English language since 1647. It was used "for all sorts of progressions from simpler beginnings."[20] According to Francisco Ayala, Darwin wanted to show that in nature life unfolds through a process that retains adaptations, suggesting not only purpose or design but also a dynamic retention in creaturely memory (see above on Bohm's understanding of movement). In other words, the living world of biology works like the universe itself: by the dynamic interplay of everything, its history, what it finds at present, its own lawfulness and chance events, while leaving the future incomplete.

Ending this section on a more general note, we can say that on all levels of inquiry, from physics to history, the sciences are now aware that the interplay of these elements seems to constitute a general array of patterns and principles responsible for the emergence of creative change over time. If this is so, it would constitute growing recognition of the notion that reality is grounded in an underlying whole, a shift David Bohm would have gladly endorsed.

Cosmic Fine-Tuning

As we saw in the previous section, it was the Judeo-Christian refusal to accept the divinity of the planets or the myth of eternally recurring cycles

---

20. Ayala, "Biological Evolution," 10.

as the basis for the world that the natural sciences developed in Western culture, at least from the time of Galileo onward. The Christian conception of the natural world differed from those of other religions and from that of classical Greece. The world was created, and therefore contingent and not necessary, but since it existed it had to have a beginning and a history. Without the awareness of the world's contingency there could be no natural science. We also noted the mysterious intelligibility of the natural world, likewise not necessary, but once discovered, it turns out to be as universal as the laws of mathematics and physics through which it becomes intelligible.

In this section, we examine another mysterious feature of the universe: the fact that its fundamental constants are extraordinarily fine-tuned, and for which there is no scientific explanation.

According to Martin Rees, British Astronomer Royal,[21] if the ratio of the strength of electromagnetism to that of gravity between a pair of protons were significantly smaller, only a small and short-lived universe could exist; similarly, the size and density of the universe are also finely tuned. If the universe were too dense, it would recollapse under the force of gravity into a singularity; if it contained too little matter, rapid expansion would occur, leaving no time for solar systems and planets to form. The constants that govern the interaction between the four fundamental forces (gravity, electromagnetism, and the strong and weak nuclear force) are likewise finely balanced. For instance, a slight variation in the strong nuclear force would have converted all hydrogen in the early universe into helium, preventing the formation of stars and water, both of which are needed to develop and sustain life. Such "coincidences" have given rise to much discussion—but also to wonderment. Stephen Hawking wrote: "The laws of science, as we know them at present, contain many fundamental numbers, like the size of the electric charge of the electron and the ratio of the masses of the proton and the electron. . . . The remarkable fact is that the values of these numbers seem to have been very finely adjusted to make possible the development of life."[22]

But there is more. Consider the following sequence of highly improbable events. The Milky Way, our home galaxy, is positioned in a

21. A highly prestigious honorary position without executive functions held by renowned astronomers appointed by the British monarch as a member of the royal household.

22. Hawking, *Brief History of Time*, 125. For a more technical article, see https://en.wikipedia.org/wiki/Fine-tuned_Universe.

medium-sized system of galaxies known as the Local Group, and this position is just as right as the position of our planet from the sun in the habitable zone. The Milky Way itself emerged at the right time. Had it belonged to an earlier group of galaxies, it would not have been able to generate the chemical elements required for the formation of terrestrial planets like Earth. The same goes for the positioning of our solar system on the outer section of a minor spiral arm of the Milky Way known as the Orion Arm (3,500 light-years across and 10,000 light-years in length), which too falls within the parameters for a habitable zone. Then there is our sun, of just the right age, neither too young (then it would be too bright) nor too old (then it would be too hot). And what about Earth? It is not only positioned at the right orbital distance from the sun but also orbits in the shadow of larger planets, the gas giants Jupiter and Saturn, blocking out the ceaseless life-threatening bombardment of comets and meteorites. Lastly, there is the moon, whose gravitational pull shortened the Earth day to create the right kind of night-and-day balance needed by plants and animals. What is so remarkable is that all later events depended for their proper functioning on earlier processes with just the right outcomes.

We can observe the same pattern elsewhere. As we saw in chapter 3, the chemical elements necessary for life, like carbon, nitrogen, oxygen, iron, phosphorus, and so on, are products of nucleosynthesis, a finely tuned process that initially operated in dying first-generation stars 10 to five billion years ago that scattered their products of heavier elements into intergalactic space.

From observations such as these, this phenomenon of fine-tuning has become known as the cosmic anthropic principle. This slightly misleading term has more to do with life generally than with human life. The strong version reads: "The universe must have these properties, which allow life to develop within it, at some stage in its history." A weaker version suggests that "the universe has not been indifferent to life."[23] Two well-known proponents of the strong version, John Barrow and Frank Tipler, have argued that "the Universe needs to be as big as it is in order to evolve just a single carbon-based life form."[24] They also claim that it "deepens our scientific understanding of the link between the inorganic and organic worlds and reveals an intimate connection between the large

---

23. Toolan, *At Home in the Cosmos*, 176.
24. Barrow and Tipler, *Anthropic Principle*, 3.

and the small-scale structure of the Universe."[25] What comes into view is that the entire range of processes at multiple levels—physical, quantum, and biochemical—as well as their emergent material manifestations are fine-tuned to an astonishing degree,[26] pointing to the recurring theme that underneath this vast creative emergence lies the movement of unbroken wholeness.

Cosmic Size Spectrum

If I suggested that in this vast cosmos the human size holds a special place, readers would shake their heads in disbelief. On the surface it would seem more likely that the opposite is true, that the human size is terrifyingly insignificant in this universe of vast time scales and billions of galaxies and stars. Didn't Carl Sagan warn against the fallacy of believing in our privileged position in the universe, calling the Earth "a lonely speck in the great enveloping cosmic dark"? Didn't he point to "our obscurity in all this vastness"?[27] Yet Joel Primack and Nancy Abrams address the problem of human insignificance from quite another angle: through the concept of size scales.

Size scales, when viewed together, form the size spectrum that encompasses all physical phenomena on all scales, from the smallest to the largest. To be sure, the laws of physics remain valid across all of them, yet they operate differently on different scales. Take gravity as an illustration. While at work everywhere in the universe, it does not work the same way everywhere. For instance, its influence is negligible at the size scale of bacteria, but it is hugely influential when it comes to the sun. In other words, in the physical world size is not something arbitrarily assigned to things. Rather, the spatial dimension or the size of a creature is part of its nature. We saw this earlier when we considered Einstein's concept of space-time. When speaking of size scales, however, it is helpful to understand that, while size is real, *size scales* help us see them in relation to other scales. Primack and Abrams call them "an intellectual zoom lens."[28] Using such a lens avoids scale confusion and ensures that each size is accorded its unique value.

25. Barrow and Tipler, *Anthropic Principle*, 4.
26. Davies, *Goldilocks Enigma*.
27. Sagan, *Pale Blue Dot*, 7.
28. Primack and Abrams, *View from the Center*, 157.

What I would like to show is that the size scale that defines the human world is the midpoint between the largest and smallest size in the universe—an astonishing assertion. However, to demonstrate it, a few words are needed about scientific measurements and notations.

The standard unit of measurement, for all sizes, is the centimeter (cm). One meter equals 100 centimeters. This can be expressed exponentially as 10 to the power of two (10^2) centimeters. Ten meters are 1,000 cm or 10^3 cm; 100 meters are 10,000 cm or 10^4 cm, and so forth (the raised number, the exponent, shows the number of zeros behind the one). Mountain heights are measured in thousands of meters. The size scale of mountains is therefore on the order of 10^5 cm.

For sizes smaller than one centimeter, the same notation rule applies, only now the exponent is negative. Take a typical cell in the human body; it is one-thousandth of one centimeter (0.001 cm) or, in exponential notation, 10^{-3} cm. The negative exponent shows the number of zeros behind the decimal point.

The difference between 10^2 and 10^3 or between 10^{-3} and 10^{-2} is, in each case, a factor of 10. Differences of a factor of 10 are so-called orders of magnitude. With these few tools in hand, we are ready to consider the cosmic size spectrum.[29]

Size in the universe is determined by the interplay between space-time and gravity and by quantum mechanics.[30] The largest visible size in the observable universe is the distance from us to the cosmic horizon (10^{28} cm). Scientists call the smallest size the Planck length (10^{-33} cm), which is the *minimum* mass that quantum mechanics allows to be confined in so tiny a region. The difference between the largest and smallest size scales—that is, between the cosmic horizon and the Planck length—is 60 orders of magnitude (10^{60}), an enormous number.[31] Table 4.1 summarizes it in steps nearest to five orders of magnitude.

---

29. Huang and Huang have created a fascinating interactive webpage that allows a dynamic exploration of all size scales in detail. See http://www.htwins.net/scale/index.html.

30. General relativity says you can only pack a certain amount of mass into a region of space-time before space collapses under the pressure of gravity. Quantum mechanics sets the minimum size limit. So-called particles like electrons and protons have extremely small masses and are always in motion. Their location is a probability function, and their size is given by the space in which they can be found with statistical confidence (the Planck length). It is the smallest possible size in the universe, 10^{-33} cm, which is also the maximum mass allowable by relativity.

31. Primack and Abramsm, *View from the Center*, 159.

Table 4.1. Cosmic Size Spectrum[32]	
Element	**Size (cm)**
Cosmic Horizon	10^{30}
Galactic Supercluster	10^{25}
Constellations / Galaxy	10^{20}
Solar System	10^{15}
Sun	10^{10}
Earth / Mountains	10^{5}
Human	10^{2}
Cell	10^{-5}
DNA Molecule	10^{-10}
Atom	10^{-15}
Atomic Nucleus	10^{-20}
Dark Matter ??	10^{-25}
Planck Length	10^{-30}

The light grey area comprises the totality of the human world—that is, the range of size scales that, until recently, was the limit of human intuition and thus the focus of all earlier cosmologies. The slightly darker area in the table shows graphically how the human size scale is indeed situated at the midpoint between the largest and the smallest size scale in the universe. This surprising observation glows with an even more mysterious luster when we realize that it does not stand in isolation but represents the *humanized* outcrop of cosmic history, showing in another way how profoundly connected we are to the evolution of the smallest—as well as the largest—structures in the universe.

Human Roots and Culture in Cosmic History

In the thought experiment that follows, we are going to compress the 13.8 billion years of cosmic history into one calendar year, the big bang at 00:00 hours on January 1 and the present moment at midnight December 31. This model will highlight the roots and antecedent stages of human evolution in cosmic history. The idea of such a calendar is not new. Carl

---

32. Data source Primack and Abwrams, *View from the Center*, 159.

Sagan proposed such an approach more than 40 years ago in his *The Dragons of Eden*.[33] On that scale, every second represents 434 years, every minute 26,000 years, every hour 1.6 million years, and every day 37.7 million years. This bird's-eye view of events that shaped our world lets us see our place in it from another angle, the dimension of time.

After the formation of our solar system, it took 30 days on our calendar for multicellular life to appear, but once it existed everything seems to have sped up: fish and amphibians appeared after another 18 days, reptiles after five more days, and mammals three days later. Next, after life appeared on land, major extinction events occurred in intervals of two to three days on our calendar, whereby each time almost all (75–95 percent) of the existing species were lost. So far, there are no scientific explanations for these extinctions.

When we consider the human journey on this scale, pondering how long it took the universe to incubate the genus *Homo*, we discover ourselves as very late arrivals among the living creatures: modern humans arrived a mere six minutes ago! Certainly, our earliest ancestors who made the first stone tools arrived a good deal earlier, but still late by cosmic standards, just one hour and 54 minutes ago. The last ice age began four minutes ago, lasting in the Northern Hemisphere until 30 seconds before the present. In the Southern Hemisphere the first prehistoric seafarers settled in Australia two minutes and 30 seconds ago.

Cultures and civilizations developed after the Neolithic Revolution, which invented agriculture—probably pioneered by women, who, as early plant and seed gatherers, knew something about the natural processes involved. It happened in the Fertile Crescent some 30 seconds ago. In the same region, as a first, Sumerian civilization appeared about 12 seconds ago on our calendar, followed by the Egyptian and Minoan civilizations soon after. Writing was invented about 10 seconds ago, which was used three seconds later to record the legal code of Hammurabi; another two seconds later bronze metallurgy was invented. In short, all recorded human history fits into the last 10 seconds.

Around the time when the kingdom of Israel was founded (seven seconds ago), a major shift in thinking occurred across the inhabited world. Great intellectual and religious systems emerged that would affect human society and culture for millennia to come, the birth of the Hebrew prophetic tradition, of Judaism and Christianity, of Buddhism,

---

33. Sagan, *Dragons of Eden*.

Confucianism, Zoroastrianism, and others. Jesus Christ appeared five seconds ago. One of the key features of this development was an innovation of great spiritual importance, *the articulation of the human striving for transcendence.* It was for this reason that the German philosopher Karl Jaspers named it the Axial or Pivotal Age; it lasted slightly less than three seconds, from 800 BCE to 200 CE.

Table 4.2. Human Roots and Culture in Cosmic History
(based on a one-year cosmic calendar)

Event	Date			
Big bang	Jan 1			
Milky Way forms	Mar 15			
Sun and solar system forms	Aug 31			
Oldest rock on Earth	Sep 16			
First life (cells without a nucleus)	Sep 21			
Photogenesis	Oct 12			
Free oxygen in Earth's atmosphere	Oct 29			
Complex cells form	Nov 9			
First multicellular organisms	Dec 1			
Simple animals (sponges, jellyfish)	Dec 14			
Fish and proto-amphibians	Dec 18			
First Extinction 86 percent Species Loss	Dec 19			
Land plants	Dec 20			
Insects and seeds	Dec 21			
Second Extinction 75 percent SPL	Dec 21			
Amphibians & Reptiles	Dec 23			
Third Extinction 95 percent SPL	Dec 24			
Early mammals	Dec 26			
Fourth Extinction 80 percent SPL	Dec 26			
Birds, flowers	Dec 28			
Fifth Extinction 76 percent SPL	Dec 29			
Primates, first frontal lobes in brain	Dec 30	Time Before Now		
Apes and hominins	Dec 31	Hrs	Min	Sec
Australopithecus	Dec 31	2	14	
Primitive humans and stone tools		1	54	
Homo habilis		1	24	

Homo erectus		46	
Modern humans		7	42
Neolithic Revolution / agriculture			28
First civilization (Sumer)			12
Axial Age begins			7
Jesus Christ			5
Axial Age ends			4
Middle Ages			2
Reformation			1
Mars Landing			Now

This experiment in time compression reveals that since the beginning more has come into existence—more of being, more of life, more events that further life like oxygenation, more of complexity, more of color, more diversity even against selection pressure, even in the face of five major extinction events. It also shows us the astonishingly deep roots of human existence in cosmic history, raising more questions at the same time. What, more does it tell us about ourselves? What meaning can Christians draw from these observations? Let me begin such reflections on a cosmological note. If we take into account an earlier observation that a positive correlation exists between the size of the universe, its age, and the state of its unfolding, right up to the appearance of humans, then this correlation is not confined to the tail end of the process, but characterizes cosmic history as a whole. John Polkinghorne makes the same point when he writes: "A universe as large as ours could have been around for the fifteen billion years it takes to evolve life—ten billion years for the first generation of stars to generate the elements that are the raw materials of life, and about a further five billion years to reap the benefits of that chemical harvest."[34]

Thinking in terms of self-conscious life, we might say that because the size of the universe is inextricably bound up with its age, it is just the right size and age to bring forth a creature of self-reflective consciousness. This means that such a life is only possible in this golden age when the universe is neither too old nor too young. Assuming the universe were only one-tenth of its present age, there would not have been enough time to produce life-promoting elements. After all, it takes at least 10 billion years for star furnaces to produce elements heavier than helium,

---

34. Polkinghorne, *Beyond Science*, 84.

particularly carbon, on which all life depends. Conversely, if the universe were 10 times older than it is, most stars, like our sun, would already have burned out to become white dwarf stars so that solar systems like ours would have vanished long ago.

A more refined look at the cosmos in calendar view reveals two further aspects that compel our attention. First, over billions of years the unfolding of the universe proceeded from less to more, from simple forms to complex ones—from helium to carbon, from single-celled life to multicellular organisms—all without miraculous interventions or breaches in the physical and chemical order. Second, life and consciousness (mind) emerged incrementally, as if to suggest that in the unfolding of the natural order something of increasing fullness or completeness was being worked out. The more arrived organically from within the creation, neither added from outside nor inserted from above. It emerged from the dynamic and self-organizing patterns inherent in what already existed. Thus, the grandiose project looks more like a work of art in the process of being whose plot is still being written than a finished product. Remarkably, human consciousness emerged as an orchestrated outcome of cosmic history to date. But if the rise of consciousness was so deeply and intimately connected to the cosmic process, is it too far-fetched to apprehend its gradual emergence as a higher-level cosmic awakening?

None of these descriptions sit well with the Christian tradition, an issue the English poet Owen Barfield, one of C. S. Lewis's closest friends, raised several decades ago. He called it "a most startling blind-spot" in Christian thought. If Christianity is highly allergic to any perception that it is a historical religion like any other and thus a product of evolution, Barfield attributes this reaction to a secret fear of committing idolatry. Yet, he argues, it was the Christian conception of linear time that differentiated Christianity from all other religions, rendering their fear groundless. A good look at the data on a timeline should have persuaded Christians long ago that the history of both the Earth and human beings belong to one coherent evolutionary process. Moreover, and more crucially for individual believers, those who believe that the incarnation of God in Jesus Christ is the climax of that history and Christ indeed its savior, they would see, if they only looked closely enough, that "we are still very near to that turning point, indeed hardly past it; that we hardly know as yet what the Incarnation means."[35]

---

35. Barfield, *Saving the Appearances*, 167–68.

Conclusion

Humanity's fascination with the starry heavens is well attested from our earliest beginnings. Already in Paleolithic times our ancestors constructed monuments of astronomical and religious significance. Religion and cosmology are two sides of the same quest: understanding our place in the cosmos and touching the ever-receding horizon of the mystery ahead of us. To demonstrate, we began this chapter with a brief survey of cosmologies that governed the worldviews of the ancient Near East. Despite differing founding stories, they shared a very similar cosmological picture based on what people at the time knew, observed, and experienced—a flat Earth with the dome of the sky above into which the celestial bodies and their courses were set. The Earth was the unrivaled geographic center of the cosmos. In classical antiquity under the influence of Plato and Aristotle, the sun and the then-known planets were thought to orbit the Earth in perfect concentric circles whose observable order testified to a superior intelligence behind all that exists. During the early Christian period Ptolemy's geometrically structured model ruled. Still guided by Plato's philosophy of ideal forms, this model pictured the planets and the sun moving in epicycles around the Earth with the Prime Mover residing above the most distant sphere. With the sanctions of the church, this cosmology endured for more than 800 years until the Renaissance, when it was overtaken by the Copernican revolution.

Its collapse made way for a new picture of the world. As the imaginary hierarchies of older cosmologies faded, the newly emerging cosmology tended to increasingly reflect the mechanisms discovered in nature. Since Earth was no longer central, humans were likewise dislodged from their central position, making it more difficult for people to experience the world as God's creation. When Newton introduced new concepts of time and space as absolutes, and space as a container, the mechanistic conception of the universe was complete, securing its three-century-long cultural and intellectual dominance in the West. Newton's God was the great engineer who had constructed a cosmic clockwork, a view that even infiltrated theology. Although God still stood behind the laws of physics, once the universe was launched, God could retreat beyond the heavens. Superfluous in creation and in history, God was reduced to a formal absentee landlord.

However, this neat scheme collapsed with one stroke of Albert Einstein's pen in 1905, when he announced his theory of special relativity.

Space and time were no longer absolutes but formed a single dimension relative to each other. This shift in physics was so groundbreaking that it demanded a new cosmology in which gravity and space-time combined to determine the large-scale nature of the universe. Discoveries in astrophysics and cosmology gave rise to previously inconceivable proposals: the universe is evolving; there are no separate independent entities as everything is connected to everything else; order and creativity are built-in properties of matter; matter and mind are inseparable; chance and lawfulness are not mutually exclusive but material manifestations of an underlying unity; the fundamental constants in the universe are finely adjusted to make life possible.

In terms of the evolution of life, we saw that humans, although late arrivals, represent another breakthrough—the birth of self-reflective consciousness. In the flesh of human brains, the very materiality of the universe became conscious of itself, suggesting that with rising complexity something ever more complete is being worked out. And a new metanarrative suggests itself: the new story in its unfolding at the cosmic level sets the pattern for all other forms of evolution, geophysical, biological, human, and sociocultural. Still in progress, this process unrolls toward a yet-to-be-actualized future and has led so far via human evolution to an incremental awakening of cosmic self-consciousness. Filtering this context through the Christian lens, the life, teaching, death, and resurrection of Jesus Christ as a turning point in history is so recent that its large-scale meaning still endures as a matter of hope for a new creation conceivable as a cosmic awakening of love.

5

Humanity—An Evolving Phenomenon

CONSISTENT WITH THE AIMS of this book to encourage Christians to befriend the findings of modern science, this chapter offers a synoptic view of the human journey from its earliest prehuman beginnings in deep history to the appearance of *Homo sapiens*. What this picture shows and what we are often unaware of is the dawning of a higher consciousness among our distant ancestors that reached beyond animalic awareness and the harsh reality of natural selection. If Christians today refuse to look at the evidence without prejudice, it is good to remember that once the Roman Catholic cardinals of Galileo's day refused to look through his telescope, yet history vindicated him whether they believed his observations or not.

Based on the premise that *Homo sapiens* is the living embodiment of our evolutionary history, the chapter closes with the proposal to embrace rather than ignore this history.

Background

A prime tenet of Christian anthropology is that humans are created in the image and likeness of God (Gen 1:27). Human beings are to be the bearers of God's image and communication, even of his self-revelation. More specifically, we have been created as God's counterpart—as God's coworkers and dialogue partners—and placed explicitly in a special situation of response-ability toward the Creator. Equipped with such lofty capacities, we are deemed to stand head and shoulders above all other creatures. Many people to this day believe that the biblical view fully explains the phenomenon of humanity: our origin and place in history, our relationship to the natural world, our unquestioned biological success, our inherent religiosity, our intelligence, our inquisitiveness, and our self-consciousness, not to mention our ability to evolve civilizations and cultures.

Yet, much to the surprise of many, ample light can be shed on the human phenomenon by studying other living creatures and even the nonliving world. Charles Darwin studied animals, plants, and geological phenomena before publishing his *On the Origin of Species*. To avoid controversy, Darwin stayed tactfully away from any reference to humanity, except by implication, and yet, as Theodosius Dobzhansky points out, "[t]he book contained perhaps the most revolutionary scientific idea of all time concerning the nature of man: man has evolved and is evolving."[1] Today, more than a century and a half after Darwin, not only has the concept of evolution become embedded in all scientific disciplines (not always correctly understood), but evolution has become part of humanity's self-image.

Yet, millions of Christians reject evolution because they believe the biblical text trumps scientific evidence. This attitude seems perplexing because Christianity, as a religious phenomenon, is inherently evolutionist, as Augustine (354–430) expressed so clearly. Doesn't Christianity claim the Old Testament as its founding revelation, rooted in Israel's earliest memories of her patriarchs Abraham, Isaac, and Jacob? Doesn't this history show the progressive drama of God's engagement with his people, their liberation from Egypt, the conquest of the promised land under Joshua through to the monarchy under Saul and David and on to its demise, culminating in Israel's Babylonian exile? And isn't the entire biblical narrative to be understood as God's story brimming with messianic promises thought to have been fulfilled in the Christ event, although it still awaits further unfolding? Isn't the trajectory of these promises openly evolutionist, not to mention that God promised the coming of Christ followed by the church age, still awaiting the arrival of God's kingdom? Against this background, one can only assume that current resistance to an evolutionary understanding even of its own history has more to do

---

1. 139 Dobzhansky, *Mankind Evolving*, xi.

with an ideological freeze than with the nature of the Christian revelation itself, as I have attempted to show in chapter 2.

On the other hand, there are many believers today who have begun to take scientific discoveries seriously. They usually begin their journey by abandoning the so-called plain reading of the Genesis text, for, in the face of overwhelming evidence, it can no longer serve as the master text of how creation happened. Moreover, this evidence points to the dynamic and process-like nature of the world, where nonlinear and chaotic systems are more common than linear ones.

As shown earlier in the book, an evolutionary view conflicts neither with the Hebrew reading of the Genesis accounts nor with Israel's root experiences of God to which their history testifies. What's more, the denial of evolutionary mechanisms at work in the unfolding of the natural order would simply continue to shroud what is really occurring. And, is it not in the perception of an antiscientific Christianity that we find one of the main reasons why young people especially are leaving the church in large numbers? On the other hand, Christians who do adopt an evolutionary view are taking an essential step toward closing the chasm of credibility that has opened between the church and its surrounding culture, at least in the West.

The gradual appearance on Earth of the genus *Homo* is supported by an enormous wealth of evidence. Although many questions remain about the stages and transitions in which it occurred, and the story gets a little blurry at times, present knowledge permits us nonetheless to draw an outline of the overall pattern and assign approximate ages to key events. The present consensus among anthropologists holds that the human line evolved in Africa, where skeletons and footprints from bipedal (upright-walking) ape-like creatures have been discovered. Their ancestors seem to have branched off from the chimpanzee line some three million years earlier. To start our exploration of the human branch on the tree of life, I invite you to follow me to the southern extension of the African Rift Valley, south of Lake Victoria.

Footprints in Tanzania

In modern Tanzania, some 3.7 million years ago volcanic eruptions had scattered huge amounts of a plaster-like substance across a vast area. After the ash had cooled, wild animals walked across the plain and left footprints typical of the African fauna one would easily recognize today—giraffes, elephants, baboons, and others. When rain fell, the plaster hardened and created perfect natural molds of the imprints. Subsequent volcanic activity covered them with more ash and so preserved them.

Some of the footprints were almost humanlike, but not quite. They were made by a small bipedal creature, not much larger than a bonobo, with a relatively large brain compared to its body size of just 1.2 meters.[2] This remarkable addition to the African fauna was given the scientific name of *Australopithecus*, or southern ape. They continued to exist over wide areas for more than two million years.

Around 500,000 years after they left their footprints, ice ages appeared in cycles of approximately 90,000 years as a result of wobbles of the Earth's axis that change orbit and tilt as it rotates around the sun (a cycle of precession takes 26,000 years). Although not severe at first, glaciation intensified over the next 1.3 million years. As weather patterns fluctuated and rainfall decreased dramatically, forests retreated. At the same time a new, more humanlike species appeared in the archaeological record. These mutants of *Australopithecine* stock are known as *Homo habilis*, the "skillful man," because of their toolmaking ability.[3] They manufactured simple stone tools, flaked bifacial pebbles that underwent little change for almost a million years. Their brain was half the size of the modern human brain. They were scavengers who competed for meat with other so-called passive carnivores of the African savannah; these were meat eaters—like hyenas and vultures—that let others do the hunting and killing. Archaeologists surmise that food sharing occurred here for the first time among them, possibly based on a rudimentary division of labor, as males began to procure meat as scavengers while females foraged for staples like nuts and plant foods. In this manner, *Homo habilis* lived guardedly for some 800,000 years alongside large predators such as lions and leopards, not infrequently succumbing to their ambush. However, from studies of our closest living primate relatives, the bonobos, we know that food sharing may not have been an exclusive invention of *Homo habilis*. Bonobos, whose anatomy is like the earlier *Australopithecus*, possess a remarkable

---

2. Bonobos live in the equatorial rainforests of the Democratic Republic of the Congo.

3. Biologists classify organisms into *species*: organisms that share the same gene pool. Species that evolved from a common ancestor are grouped together in subgroups above the level of species, called *genus*. From this method of biological classification, we get the two names *Homo* (genus) *habilis* (species).

range of human traits. After all, they share 98.5 percent of our DNA. Often called pygmy chimps, their social behavior differs considerably from that of chimpanzees, whose society is male dominated, competitive, and violent. Bonobos, by contrast, live harmoniously in matriarchal groups, where food sharing is the norm. Genetically, both chimpanzees and bonobos are more closely related to us than to other great apes such as the gorillas.

Homo habilis is regarded as the first progenitor of the genus *Homo*, which had emerged from australopithecine stock 1.8—2.3 million years ago.[4] But they were very dissimilar in stature, appearance, and brain capacity to modern humans; they were short, featured apelike long arms, and had a cranial volume less than half that of *Homo sapiens*. Yet, wherever their fossil remains have been found in Tanzania and Kenya, primitive stone tools are also present. Whether *Homo habilis* is a direct ancestor of *Homo erectus* is still debated by paleoanthropologists. Recent discoveries indicate that both species existed side by side for nearly 500,000 years.[5]

Homo Erectus and Their Relatives

This new species arose in Africa around two million years ago. *Homo erectus* or "upright man" survived for almost two million years and is regarded as the most durable of all the human species to date. Their advent represents a major step toward modern human stature. With time, they and their relatives spread over a vast geographic expanse from South Africa to central Europe to northern China and Java. As the name implies, this species was habitually bipedal and featured modern body proportions, so that they are assumed to have started the evolutionary impetus that eventually led to *Homo sapiens*.[6]

Technologically, their chief characteristic was the hand axe, a pointed stone tool whose rounded top fitted neatly into the palm of a human hand, possibly modeled after the sabertooth of a well-known feline predator. The evolution of animals speeded up in parallel, lasting for a million years. The open grasslands had earlier led to the appearance of sleeker grazing animals that replaced rhino predominance. This new

---

4. Shurr, "When Did We Become Human?," esp. 50.
5. Shurr, "When Did We Become Human?," 51.
6. For a detailed exposition see Dobzhansky, *Mankind Evolving*, 174–83.

wave of evolutionary innovation added such animals as bison, sheep, and wild hogs.

Our *Homo erectus* ancestors were taller and had larger brains than *Homo habilis*, who endured for only a few hundred thousand years alongside *Homo erectus*. The latter were no longer scavengers but hunters of large prey and gatherers of plant foods. They also used fire purposefully from about 1.4 million years ago.[7] There is evidence of routine cannibalism for food. But after *Homo erectus* had mastered fire there was little change in the life of these ancestors for a million years. As they had little impact on animal evolution, it went on undisturbed in all parts of the globe.

Around 800,000 years ago, more challenging climatic conditions pushed human life into new ecological niches. Originally stable during the warmer periods, human existence was forced to adapt to alternating warmer and colder conditions as a series of ice ages of increasing severity struck their habitat.

The breakout of *Homo erectus* into Europe and Asia remains one of the most remarkable events in *Homo*'s history. The best early evidence comes to us from a site in Caucasian Georgia, Dmanisi, where a cache of hominid fossils was found dating to around 1.8 million years ago. However, they were of shorter stature, their braincases were smaller, and they used cruder stone tools than their African relatives. These facts have led to questions about their ability to engage in long-distance movements. Other less well-documented fossil finds exist in Pakistan, dated to 1.6 million years ago, and in Israel, dated to 1.4 million years ago. *Homo erectus* seems to have entered Europe around the same time. There are finds in northern Spain (between 1.4 and 1.2 million years ago) and Italy (900,000 years ago). A younger specimen from northern Spain (dated to 780,000 years ago) has been identified as *Homo antecessor*, considered to be the stem species that gave rise on the one hand to Neanderthals and on the other to *Homo sapiens*.

In Europe too, around 600,000 years ago climatic fluctuations caused by the Gulf Stream in the north and the Alps in the south produced new pressures to adapt; the same was true in freezing northern China (Choukoutien), where *Homo erectus* hunted large game like rhino, bison, and horse. Finds of skulls with purposely enlarged foramen suggest that the protein-rich brain was extracted for dietary or possibly cultic reasons.

---

7. James, "Hominid Use of Fire."

The Earliest Cosmopolitan Homo and Their Successors

The exodus from Africa must not be understood as a single event. Certainly, the fossil record shows genus *Homo* followed the typical pattern of local and regional diversification on the home continent (characteristic of successful mammal species). At the same time, an ancestral drive for "globalization" showed up long before the appearance of modern humans.

The line that would eventually give rise to *Homo sapiens* featured another ancestral link, *Homo heidelbergensis*. These archaic humans appeared in the later phases of *Homo erectus*, around 600,000 years ago, also in Africa, where they evolved further and from where they spread to Europe.[8] It was first described from a 500,000-year-old find in Germany, near Heidelberg. This species has become well known from other finds since: a 600,000-year-old find in Ethiopia, as well as fossils discovered in France, Greece, Zambia, and China, some of which were more recent (dating from 200,000 years ago).[9] Despite some variations in detail, facial morphology remained remarkably consistent. In France, anthropologists discovered the remains of a 350,000-year-old beach shelter that featured a hearth containing charred animal bones. More evidence for the controlled use of fire comes from a 595,000-year-old site in Israel, although the *routine* use of domestic fire is not attested until 400,000 years ago, from the Terra Amata site near Nice in France. Several carefully crafted wooden spears date from the same period, suggesting an advanced form of ambush hunting. Undoubtedly, compared with his relatives, we meet in *Homo heidelbergensis* a cognitively more complex variety of the genus *Homo*. Overall, the evidence suggests a picture of substantial biological diversity, although signs of cultural advance are still lacking, and the use of symbols in the production of cultural artifacts was still more than a quarter of a million years away.

Homo heidelbergensis is an ancestor of the Neanderthals who lived in Europe, in the Middle East, and in Western Asia from about 200,000 to 28,000 years ago, and also of modern humans. Today, Neanderthals are considered cousins (no longer parents) of *Homo sapiens*. The present consensus among anthropologists holds to a single African origin of *Homo*. On this basis, anatomically modern humans (AMH)—not yet *Homo sapiens*—arose in Africa around 200,000 years ago, spreading out

---

8. Shurr, "When Did We Become Human?," 52.

9. McHenry, "Homo heidelbergensis," https://britannica.com/topic/Homo-heidelbergensis.

from there into Europe and Asia about 70,000–50,000 years ago, where they replaced Neanderthals and *Homo erectus* through extinction and interbreeding. Aboriginals had already settled in the arid regions of Australia in the Flinders Ranges of South Australia around 49,000 years ago,[10] having first arrived 20,000 years earlier on Australia's north coast.

While a leap in stone tool technology had occurred with *Homo erectus*, who invented more refined flaking techniques for use in axes, scrapers, and points, this industry was taken to a new level with *Homo heidelbergensis*, who was able to produce custom-shaped flakes. By the Mesolithic in Eurasia and northern Africa (about 250,000–200,000 years ago), blade-making technology had advanced to produce elongated flakes that could be used as blades with sharpened cutting edges, some shafted with wooden handles. It took another 100,000 years before thin-blade technology appeared, together with the routine use of bone, ivory, and antler in toolmaking and in articles for personal adornment. This revolution of the Upper Paleolithic (about 70,000 years ago) is considered by some archaeologists an unambiguous marker signaling the rise of fully modern humans.[11]

When placed on a timeline, evolutionary advance in brain size occurred from *Homo habilis* (500–600 cm^3) to *Homo erectus* (850–1100 cm^3) to *Homo heidelbergensis* (1100–1400 cm^3) to Neanderthals (1200–1900 cm^3) to modern humans (950–1800 cm^3). This is not to suggest that human evolution occurred in a straight line as if *Homo habilis* begat *Homo erectus* and Neanderthals begat modern *Homo sapiens* and so forth. As mentioned, Neanderthals are our cousins, not our parents. Similarly, our ancient ancestors are not simply more primitive models of ourselves; rather, the Earth was home to several human species at any one time from about one million to about 10,000 years ago.

Until recently, the place of genus *Homo* in the food chain was somewhere in the middle. As hunters and gatherers for 1.5 million years, they hunted small animals while remaining prey for larger ones. It was only in the last 400,000 years that some of our species began to hunt large game, jumping eventually to the top of the food chain. Ecologically, the consequences of this leap from prey to top predator are still with us, and

---

10. Hatfield, "Introduction," esp. 9. Also see recent reports of finds at the Warratyi rock shelter in the Flinders Ranges, 550 km north of Adelaide, Australia. http://www.abc.net.au/news/2016-11-03/rock-shelter-shows-early-aboriginal-settlement-in-arid-australia/7983864.

11. Hatfield, "Introduction," 12.

the impact grows worse year by year, as the planetary ecosystem never had time to catch up with our ever-advancing hunting and killing methods. Our ancestors first began to cook food about 700,000 years ago, but only with the appearance of the Neanderthals were animal skins used for clothing. Naturally, this raises the question, "What makes us quintessentially human?" Before addressing it, let us first trace the emergence of *Homo sapiens* through the pages of archaeological history.

The Rise of Homo Sapiens

Two million years of innovative thrusts in the genus *Homo* prepared the ground for a new creature to emerge from the dimly lit past, *Homo sapiens*, the self-styled "wise man." In a fraction of that time, our species has grown so dominant that it now threatens the entire ecosystem of the planet and, with it, its own existence. To describe its appearance, paleoanthropologists find it helpful to distinguish between *anatomically* modern humans and *cognitively* modern humans. While chapter 6 explores the emergence of human intelligence and the mystery of the human mind, we focus here on species-specific traits that have puzzled evolutionary theorists in several ways. The question of where these creatures had come from and when is puzzling, to be sure; even more perplexing is the question of their unprecedented and extraordinary technological ascent. Until their appearance, the pace of hominid evolution in terms of both anatomy and behavior was exceedingly slow. While this snail's pace seems to suggest that genes need relatively stable conditions over long periods to express themselves, the preposterous speed of cultural developments over the last 10,000 years—a mere nanosecond in evolutionary terms—thoroughly contradicts this idea. What is it about this creature that in a twinkling of an eye has transformed the world?

Given this distinction, it is not surprising that our astonishing cognitive powers are regarded as the one quality that most strikingly distinguishes us from all other species on the planet. By this is meant the ability to symbolize, which allows us to gain both an objective and subjective grasp of the world around us, other human beings included. Although an evolved feature of our neuronal wiring, how and when we acquired this unusual capacity is still open to debate. What is certain, however, is that it enables us to remember the past and imagine the future, including alternative possibilities. Certainly, growth in brain size had something to do with it, and it cannot be doubted that among the variegated branches of human emergence we are the only hominid line that, by the Upper Paleolithic, can lay claim to symbolic cognition. I will have more to say about this feature when we explore the emergence of human intelligence and the mystery of consciousness; for now I'll confine myself to a few words on the evidence for symbolizing behavior.

The best place to start is with a cave 300 kilometers east of Cape Town, 35 meters above sea level and 100 meters from the shoreline, the Blombos Cave.[12] Today it is *the* archaeological site where fieldwork has yielded striking evidence for the creative explosion that went hand in hand with the emergence of early modern humans, turning earlier views on their head. After the discovery in 1940 of stunning cave art in southern France dating from 17,500 years ago, archaeologists believed that modern behavior like fishing, the development of bone implements, the use of personal ornaments, as well as symbolic cognition—that is, the ability to identify and make figurative representations of things—did not evolve until about 50,000 years ago. Now, since the 1990s, with excavations at Blombos Cave and at other sites in South Africa, paradigm-shifting evidence has accumulated, demonstrating that the creative phase of the human mind began much earlier than previously thought, somewhere around 100,000 years ago, for it was already flourishing around 70,000 years ago. This evidence has been central to the debate about identifying the time and place of our cognitive and cultural beginnings. At Blombos Cave, over millennia of occupation, these ancestors left behind diverse sets of archaeological footprints showing they engaged in multistep stone toolmaking that combined bone and stone in new tools, engraved bone and used ocher (processing pots have been found), manufactured marine shell beads, adopted new dietary habits discernible from the remains of shellfish, birds, tortoises, ostrich eggshells, and more. All this has led to a complete rethinking of the timing and location of the cognitive leap that propelled our species into unforeseeable creative possibilities.[13] In sum, the South African coast may be regarded as the cradle of the human mind.

12. Henshilwood et al., "Early Bone Tool Industry."

13. Some scholars still hold to a later development of symbolic behavior (i.e., from 50,000 years ago), arguing that the necessary neural capacity was not visible before that time in bony structures, while it was far more pronounced in Cro-Magnons—the later European variety of *Homo sapiens*. See Tattersall and Schwartz, "Evolution of the Genus *Homo*."

When *Homo sapiens* appeared, several other hominids existed at the same time—*Homo neanderthalensis*, *Homo erectus*, and, on the Indonesian island of Flores, a dwarfed variety, Homo floresiensis, who lived at Liang Bua from at least 190,000 to 50,000 years ago. The pages of archaeological history also show that the arrival of *Homo sapiens* and their incursion into the territory of other hominid species went hand in hand with the rapid extinction of the other species.

In monumental cave art, we find highly potent testimony to the formidable and unprecedented capacity of *Homo sapiens* to make use of symbols. This and other cognitive faculties would soon cause this species to become the dominant species on the planet, for good or ill.

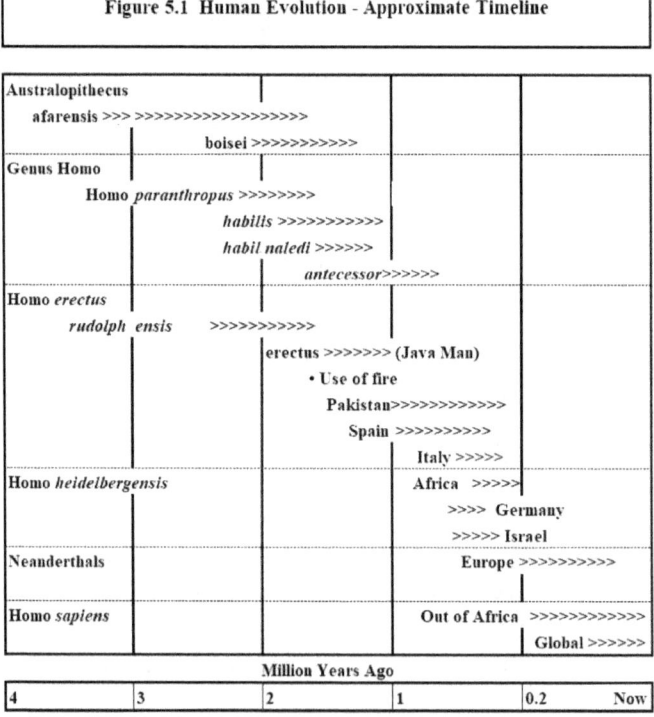

Figure 5.1 Human Evolution - Approximate Timeline

Sources: *Encyclopedia Brittanica*; Smithsonian Institution.

Developing Skills in Human Infancy

It is popular among paleoanthropologists to assume that early humans were not hunters, at least not initially, but scavengers who used primitive tools to scrape meat off large animal bones or break them to extract the nutritious marrow. What is certain is that our early human ancestors did create stone tools and passed on the technique to succeeding generations for almost a million years with little change in design. They used a method known as knapping. They struck a suitable stone with another in a specific way to create stone flakes with sharp edges, then used them to butcher the meat they were able to snatch from kill sites of predators or to dig for grubs and edible roots. All this activity demanded a high level of manual dexterity as well as cognitive abilities that enabled them to recall the how-to of these methods.

To be sure, rudimentary forms of this behavior are observable in the animal kingdom. For instance, when cracking nuts with stones, adult chimpanzees are able to use just the right amount of force to crack the shell without smashing the nut. Youngsters need much practice under the scrutiny of their mothers (for up to seven years) before they master the technique. With *Homo habilis* we see a creature that breached the confines of the animal world for the first time. While chimpanzees and certain other animals are known to use simple tools and even make them, the ability of *Homo habilis* seems to have worked at a higher level of cognition. For instance, they were able to pick the right kinds of stones for knapping and must have possessed more fully developed manual dexterity, for toolmaking rose significantly above animal attainment. The earliest reliably dated artifacts using the technique of knapping are 2.3 million years old, showing finished edges that cannot be manufactured by simply banging rocks together. Besides, knapping is a learned technique where the second round of strikes must sharpen the edge, not blunt it.[14]

If you had asked sociobiologists and evolutionary psychologists 20 years ago about the way humans had developed their cognitive capacities, they would have pointed to the universal phenomenon of "human nature," speculating that it had laid down the basic structures between 200,000 and 30,000 years ago that have remained largely unchanged since then. However, this static model is no longer coherent. Studies of intellectual and emotional growth, even among early primates, have shown that the capacities that distinguish us from the animal world (symbolic

---

14. Donald, "Mimesis Theory Re-Examined," esp. 178.

and logical reasoning) developed gradually and earlier than generally assumed.[15] The cognitive capacities for pattern recognition and symbolic thinking emerged, according to Stanley Greenspan and Stuart Shanker, at the rich interface between biological and emotional developments, having been transmitted from one generation to the next by caregivers during early infancy even among early primates. For instance, baby marmosets and tamarins engage in intense interchanges with their mothers—mutual gazing, and looking at their faces, accompanied by much vocalizing between them. Rhesus babies snuggle close to their mothers, who gently rock them while both engage in much mutual eye contact. Intense gazing is accompanied by vocalizations and other forms of emotional signaling that even increase as the infant grows.

The same may be said for baboons and other mammals. Here, the typical two-way emotional interaction between caregiver and infant is more purposefully directed, while in chimps we can observe subtle anticipatory skills, especially when complex social problems must be solved. For example, during hunting parties chimpanzees coordinate their moves through sophisticated emotional signaling as they anticipate how other members of the group will move as well as the movements of the prey. Paleoanthropologists believe it is highly probable that this kind of behavior was also available to early human ancestors as far back as the australopithecines more than five million years ago.

When archaic *Homo sapiens* decorated their bodies with pigment and engaged in the artistry of cave painting between 600,000 and 60,000 years ago, they proficiently used a form of presymbolic communication and may even have expressed abstract thoughts and emotional states this way. Around 70,000 to 60,000 years ago *Homo sapiens* engaged in sea voyages (e.g., settling in Australia); they not only engaged in complex group behavior (planning and decision making), but they also correctly perceived the link between natural events and group decisions, which enabled them to act accordingly.

When considering the development of new cognitive capabilities in our prehuman ancestors, the fact that they were social creatures long before the arrival of *Homo sapiens* must not be overlooked. After all, these capacities do not exist at the level of biology alone. While natural selection may explain the emergence of some basic biological traits, it is unable to decode the more complex processes we engage in such

---

15. Greenspan and Shanker, *First Idea*, 92–95.

as symbolizing and communication. The higher reflective skills necessary to conduct increasingly more complex social interactions cannot be provided by fixed biological pathways. We also saw that the human mind evolved predominantly during the early stages of infancy through mother-infant interactions. In short, the higher capacities associated with mind are both genetically determined and culturally transmitted.

Irrespective of whether these are prehuman, archaic, or modern, ongoing early interactions with caregivers will teach offspring how to focus attention, give shape to desires, and manage intentions. The same holds true for the species as a whole; higher capacities of that kind can only emerge in infants when they are nurtured by affectively engaged caregivers who pass on their practices to succeeding generations.

The Dawning of an Interior Horizon

As we saw above, higher forms of cognition play a decisive role in integrating observations and experiences more comprehensively than what the perception of sense data alone would allow. For instance, the experience of being enthralled by a beautiful sunset requires integration at a higher level than optical impressions. When our ancient ancestors stood mesmerized by the flamboyant display of colors in the heavens, their experience, just like ours, was derived from this higher form of representation. In such moments they, and many generations since, have known the sun's warmth, encompassing them with its bounty. They would have felt its inexplicable energy as a source of well-being for all creatures, mysteriously causing everything to grow and prosper. Or, during a devastating drought, they would have experienced the sun's unrelenting heat that drove them to the brink of endurance, confronting them with life's fragility, finiteness, and mortality. This existential reality, which is the result of our mind's ability to carry out such higher-level integration, is not foreign to us, and to the extent that our ancestors possessed the same brain, they would have experienced it the way we do.

When survival is at stake, having access to modern scientific data about the sun as a gigantic mass of high-temperature hydrogen and helium is of little help. While rationalism easily dismisses the experiential-symbolic dimension of human existence, its relevance must not be ignored. There is simply too much evidence that even some of our primate ancestors reached for a horizon beyond the merely factual and tangible;

modern chimpanzees possess this capacity, which is the ability to identify empathically with the experience of others. Famous primatologist Jane Goodall reported how among a group of chimpanzees one particularly high-ranking ten-year-old female had died from wounds inflicted by a leopard. Six males and six females gathered respectfully around her dead body. Only the group's older females were permitted to inspect the victim's wounds, while her younger brother alone was permitted to groom her coat for a few seconds and look at her face. All the others were driven away. After a few hours, the group slowly and hesitatingly left the scene.[16]

Judging from the solemn mood in which these chimpanzees beheld their dead companion, it may be suggested that they shared for a short time not only in the loss of a revered group member, but also in something deeper that transcended the biological fact that one was no longer alive. In these chimpanzees a new, interior horizon had dawned. The archaeological record suggests that some of our hominin ancestors perceived with greater intensity and clarity than chimpanzees not only the existence of this horizon but that it pervades all things.

The most striking cultural evidence comes from Spain. At an exceptional site in the caves of the Atapuerca Mountains, which has been occupied continuously for 800,000 years right up to the present, many archaic and modern human fossils have been discovered.[17] Of special interest are skeletal remains from the Sima de los Huesos, which is believed to be the earliest recorded mortuary site. It contains the remains of many adolescents and young adults identified as *Homo erectus*. Among them is the skull of a five- to eight-year-old child who must have had a serious birth defect called craniosynostosis. One or more of the joints between the bones of a baby's skull close before the brain is fully formed.[18] As the child grows, the brain cannot develop normally, deforming the baby's head and restricting mental development. Remarkably, the child lived for at least five years, which means its family would have cared for it as it would have for a healthy child. A similar example comes from a famous Neanderthal skeleton found in France: "The "Old Man" of La Chapelle-aux-Saints. The man lived 60,000 years ago and his remains show many missing teeth as well as signs of crippling arthritis. Similar cases have been found in Iraq (Shanidar), where individuals incapable of surviving on their own

16. Goodall, *Chimpanzees of Gombe*.

17. It is a UNESCO World Heritage Site. For a detailed description see http://whc.unesco.org/en/list/989.

18. Today, this condition can be treated surgically soon after birth.

managed to reach old age. Recent reanalysis of the 50,000-year-old skull of a Neanderthal male showed that the individual had not only sustained multiple injuries but also was profoundly deaf. Yet he had lived well into his forties, an astonishing age in the Paleolithic and impossible without sustained social support.[19]

In each case, other members of the group or horde seem to have recognized in their disabled companions someone whose existence was meaningful, even precious, beyond mere biological reality. It was their ability to identify with them that perceived this interior, invisible horizon. The dawning of this horizon must have provided the emotional strength to persist over countless generations with valuing these "unnatural" ideals that did not deliver any tangible survival benefits. This trait, argues anthropologist Nancy Tanner, has been attributed to the close bond between mother and child.[20] Viewing it as a mere abstraction is one thing; trying to understand what our ancestors saw requires that we touch on a deeper question. What was it that now caused them to act against the pressure of natural selection to abandon a disabled child or a toothless grandfather? What breakthrough must have occurred that prompted *Homo erectus* parents to look after a disabled child and Neanderthals to engage in aged care?

What is at stake in these questions is something science qua science cannot adequately address, for questions of this kind touch on the great taboo of modern scientific inquiry: subjectivity. The *Homo erectus* parents and the Neanderthal horde went beyond the natural survival needs of the group because they had had a personal, subjective experience of the other. Each group had identified with the inside story of the disabled child or toothless old man, and the emerging human mind had begun to introduce into their existence a new depth to feelings of sympathy to which earlier ancestral creatures might have been oblivious. To be sure, physically, chemically, and biologically everything continued as before. Metabolisms continued the same as before, yet on the horizon of the cosmic drama something of extraordinary significance had occurred: the

---

19. https://tecake.in/news/science/neanderthals-disabilities-survived-help-social-support-39391.html.

20. Tanner, *On Becoming Human*. From evidence of this nature, German biologist and theologian Ulrich Lücke has inferred that it is not our symbolic perception alone—not even our vocal gestures or our social organization or the ability to invent and use ever more sophisticated tools (or weapons)—that makes us human, but our consciousness of a transcendent horizon (Lücke, "Der Mensch," esp. 79.

ability to perceive an inside story had dawned within the whole material structure of the universe. In mysterious silence and without fanfare, the natural world had gradually been raised to a fuller level of existence, so that from then on everything was going to be different: in the mind of some of its creatures, and here ever so dimly, the universe was awakening to its own inside story as if from a long and necessarily dreamless sleep. We deepen this understanding in the next chapter.

One of the aims of this chapter has been to point to the overwhelming paleontological evidence in favor of an evolutionary origin of the human species. This evidence is massive even without reference to genetics.[21] No doubt, many Christians have in one way or another been exposed to it during their lifetime based on other popular or scientific sources, at least in broad outline. As I have tried to show, from this evidence a picture emerges that tells its own story. *Homo sapiens* appeared at the end of a long chain of gradual evolutionary developments over two million years, sharing an ancestry with *Homo erectus*, *Homo heidelbergensis*, *Homo floresiensis*, Neanderthals, Denisovans, and Cro-Magnons.

Why We Cannot Ignore the Past

The awesome actuality of *cosmic self-awareness* realized in us humans on a tiny planet, in an insignificant corner of the cosmos, will occupy us more fully in the next two chapters. This closing section of the chapter on human evolution is devoted to an often-unrecognized feature of this reality—namely, that we, the bearers of cosmic self-awareness, are also the undeniable product of deep history.

Constituted by the residue of 10-billion-year-old exploding stars, we also carry in us the 1.5-billion-year-old attributes of the first nucleated cells; the genetic material of 400-million-year-old fish crawling on land and becoming amphibians; of 65-million-year-old large lizards growing feathers, taking to the air and becoming birds; of 50-million-year-old carnivorous mammals returning to the sea and becoming dolphins; and of primates leaving the rainforests and becoming bipedal scavengers of the African savannah, whose later cousins pole-vaulted to the top of the food chain 400,000 years ago by turning fire-hardened sticks into

---

21. Readers interested in the genetic aspects of the human story from a Christian perspective are encouraged to read a specialist's account: Janssen, *Standing on the Shoulders of Giants*.

weapons, and eventually rising to develop speech and sophisticated levels of symbolic thought.

In a manner of speaking, this is the modern creation story. Having said this, I do not wish to imply that the new story can or should rob the ancient story of its power. To the contrary. As we will see, even the evolutionary story comprises the elements at work in the biblical garden story, perceived by the ancient Hebrew intuition as the voice of the serpent, namely, the temptation to follow our lower drives, as well as the inner conflicts in trying to resist, and our fall when we succumb. In short, this biblical narrative may have a substantial basis in biology. Even so, this notion does not answer the pivotal question of how to explain coherently the Christian doctrine of a human fall that took the entire creation with it, given the very gradual awakening of the genus *Homo* to higher consciousness over hundreds of thousands of years, not to mention an even longer prehuman history.

More than half a century ago, American physician and neuroscientist Paul MacLean (1913–2007) pioneered the idea that the development of the human brain recapitulated our evolutionary history, comprising reptilian, paleomammalian, and neomammalian brain structures, to which subsequent research added the later-evolved prefrontal lobes. We will look at specific brain functions more closely in the next chapter; what I want to draw attention to first is that powerful automatic and hard-to-control propensities are still part of us. For instance, the fight/flight/freeze response to threats or trauma is an involuntary reaction of the *reptilian brain*. Under certain conditions, this evolutionary legacy may react so powerfully that attempts of other brain regions to override it may fail. Indeed, the reptilian brain is the most deep-seated of our cerebral structures, which explains why its responses lie behind our most basic and ancient drives for food, territory, and procreation. When individuals or groups are threatened or traumatized, it is this structure that predisposes them to react strongly with aggression and defensiveness.

The evolution of the *limbic system* or the old mammalian brain allowed mammals a wider and more subtle range of behaviors, including experiential learning, than what reptiles are able to deliver. It comprises such brain structures as the amygdala, hippocampus, thalamus, hypothalamus, and insula. Without going into detail about their individual contributions, the limbic system manages the emotional and dream states without which mammals cannot live. Since the limbic system has the power to generate a multiplicity of emotional states ranging from the provision of milk for the young to bonding behaviors and group adaptations, it is also capable of generating the emotional charge behind bruising rivalries and the striving for dominance, as well as addictive cravings for substances that generate feel-good responses in the brain.

The neomammalian brain or *neocortex* is a later evolutionary structure shared by primates and dolphins; in humans it enables the capacity for storing active memories, for rational thought, and for creating imaginative plans by which to test, analyze, and select future actions based on preset values or imagined benefits. The neocortex, too, is the place where subjective sensory perceptions are transformed into experience and learning, the juncture for mystical activity, spiritual aspirations, and the lure of the future. The neocortex is also capable of facilitating choices between competing drives by comparing them. There is a downside to this highly geared function: its hyperactivity. The neomammalian brain, being constantly engaged in the processing of vast volumes of internal and external data, has been dubbed by practitioners of meditation the "monkey mind" because of its incessant chatter and the tendency to leap at random from one mental event to another. It tosses countless thought fragments into its inner space without much conscious focus unless disciplined by the need for immediate action.

This ravenous behavior receives special attention in the next chapter. What comes to our aid is our *frontal lobe*, which performs the executive functions of the brain. Through it we can choose between mammalian and reptilian drives, exercise intentionality and purposefulness, and engage complex decision making. In this brain region ideas are generated, plans are formulated and scrutinized. The frontal lobe is the seat of our self-awareness and consciousness, which is more generally supported by the brain as a whole.

From such a perspective on our evolutionary past, while far from explaining everything about us, we can infer how profoundly we have been shaped by our ancestry and the cultural context into which we were born, as well as how we have developed through our choices made along the way. We also begin to see a little more clearly that, despite its vast and complex computational capacity, the human brain functions as a whole, relying on and in continuity with its reptilian, paleomammalian, and neomammalian structures, whose influence on our behavior and choices cannot be denied. As a result, we find ourselves to be a highly inventive creature, yet one ill adapted to its own high-tech culture, a creature who is self-conscious but fails to know what and why it feels the way it does. We are often self-deceived and despite our lofty aspirations; our inherited proclivities—if left unacknowledged and unaddressed—tend to generate some extremely troublesome energies, with the potential to threaten the collective well-being of the entire species. The point is that we must begin with the reality of where we have come from, individually and as a species. History tells us that formulating a set of moral ideals and freezing them in a code of conduct is not enough when we must deal with unwanted drives within us inherited from our evolutionary past.

Conclusion

I understand why many Christians might instinctively balk at the evolutionary anthropology outlined in this chapter. But bear with me. I hope to show that there are some life-giving lessons to be learned. What if, as intimated in the introduction, God—in the supreme wholeness of his being—inspired humans in our time to search, ask new questions, and share ideas on an unprecedented scale? Suppose further that despite our inability to grasp the full meaning of the information at our disposal, we had nevertheless reached way beyond the limitations of knowledge of previous generations and arrived at the kind of conclusions I have described above, with a tolerable margin of error, as a form of public revelation. If this were granted, wouldn't it make sense for Christianity to translate its doctrines into a language that explained our flawed humanness in evolutionary terms, which is to say from the standpoint of an incomplete cosmos? Wouldn't it make sense to explain our conflicting inner drives in terms of our evolutionary past and our higher aspirations as the divine lure of a yet-to-be-realized future awaiting our fuller participation? If, in view of cosmic history and our place in it, our evolved brains are indeed the seat of *cosmic self-consciousness*, then who is to say that the Creator of an emerging universe would have it any other way?

6

The Human Brain and the Rise of Human Consciousness

THIS CHAPTER DEEPENS THE quest for understanding human consciousness by exploring the human brain as the biological substrate of spirituality. Relying on the findings of modern brain research, the chapter looks at the brain's role in how we perceive God, the spirituality of children, and the workings of the mystical mind. It closes with the conclusion that the very stuff of the universe has become conscious of itself and that thus human consciousness must be understood as the inside story of the cosmos and as the seat of religious revelation in the omnipresence of God.

Toward a Biology of the Human Spirit

The development of symbolizing and higher forms of cognition had been very slow in coming. Our early ancestor, *Homo erectus*, who lived from 1.8 million to 50,000 years ago and was the longest-lived species of the genus *Homo*, relied predominantly on the immediate inputs and stimuli of what was adaptively useful to them. It was the absence of a collective working memory that permitted only a slow transformation of the brain toward higher cognitive faculties. For instance, it took 750,000 years to adapt to colder climates and 500,000 years to domesticate fire. Yet, with the late *Homo erectus* the first major transition occurred, leading to the spread of improved tools. Their production was accompanied by iconic representations, and some bore symbolic markers. One of the oldest pieces of evidence for symbol use is a lower jawbone of a wild horse marked with two oblique crosses. This bone fragment was discovered at Wyhlen in Germany, near Lörrach (Baden-Württemberg), and dates from the Riss glacial period, around 200,000 years ago. The heavy markings were made with a stone tool. Perhaps the hunter who killed the animal was proud of his success, marking the bone with a double signature of possession. Whatever his motivation, it must have met a higher-level need. German scholar of religion Georg Baudler interprets this subjective gesture as possibly a first attempt to answer one of the most fundamental questions of human existence: "Who am I?"—a point to which I will return.[1]

Modern neuroscience and allied disciplines have long speculated about the emergence of human symbolic intelligence. Some looked to enlarged brain size, increased convolution of brain tissue, and other neuroanatomical changes for primary clues; others attributed the rise of such capacities to genetic mutations, assuming these were coded directly for the brain. By contrast, American neurologist Fred Previc placed the weight on expanding neurochemical systems, and I will follow his argument.[2] He reasons that the starting point for the spurt in human intelligence was adaptive stress and subsequent metabolic changes, which, in turn, prompted the expansion of brain circuitry. These circuits depended for their effectiveness on an increased supply of dopamine (DA), a key neurotransmitter principally involved in regulating cognitive capacities in animals and humans, yet with a crucial twist: dopamine also regulates responses to thermal stress.[3]

The relevance of this argument becomes clear when seen against the background of the frequent, often drastic climatic oscillations that occurred during the two-million-odd years of human evolution. Wet periods alternated with long spells of drought, and tropical phases with icy periods of glaciation. During the outgoing Pliocene (5.3—2.6 million

---

1. Baudler, *Das Kreuz*, 34.
2. Previc, "Dopamine and Human Intelligence."
3. Neurotransmitters are also understood as the brain's chemical messengers. They transmit signals from one nerve ending to another across a synaptic gap. Transmitters are stored below the membranes of the nerve terminals; when released into the gap for short periods, they will permit a signal to travel across, causing a "postsynaptic" response on the other end. At least 100 such chemicals are known to exist. These neurochemicals are readily synthesized in the body from other substances like amino acids.

years ago), the climate in East Africa became more arid. Retreating forests turned wooded regions into open savannahs, forcing our ancestors to become fully terrestrial. This new way of life placed severe adaptive demands on physical endurance, especially in the procurement of food and water. As our ancestors were required to travel long distances,[4] adaptations for improved regulation of body temperature were needed, and these depended critically on dopamine-charged mechanisms.[5]

According to Previc, their bodily response to physical endurance pressure led in the human brain to an expansion of DA mechanisms, which at the same time enhanced their cognitive capacities. For instance, the prefrontal cortex, parts of the limbic system, the left hemisphere of the brain, as well as neural reward structures rely for their effective functioning on dopamine pathways. Not only are these regions deeply involved in managing key functions of advanced intelligence and cognition such as organized behavior, emotions, working memory, executive functions, parallel processing, spatial and temporal abstraction, motor programming, and language ability, but the ability to adjust one's cognitive and motor strategies in response to new information also critically depends on adequate dopamine concentrations. Conversely, DA depletion deactivates key brain regions, leading to serious disorders like depression, attention deficits, language deficiencies, schizophrenia, and loss of cognitive and social competencies.

This prompts the question of how the brains of our ancestors achieved a higher DA turnover, which would have fostered the emergence of higher intellectual capacities and abstract reasoning. Two factors are thought to have played crucial roles. One that I have mentioned already is the involvement of DA in regulating body temperature.[6] To prevent hyperthermia when exposed to higher temperatures and physical exertion (e.g., in traveling long distances and possibly engaging in chase hunting),[7] our ancestors needed thermoregulatory mechanisms more advanced than those of the great apes (modern apes travel less than one kilometer per day). In other words, the pressure of their new environment in the open

---

4. Previc, "Dopamine and Human Intelligence," 319–23.

5. Previc, "Dopamine and Human Intelligence," 304.

6. Lee et al., "Dopamine and Thermoregulation," 589–98; also Previc, "Dopamine and Human Intelligence," 321.

7. Chase hunting was a technique used by early hunters, who lacked effective weapons. These hunters chased an animal until it succumbed to heat exhaustion before they did.

savannah demanded large releases of DA, which in turn prompted the coevolutionary effect of enlarging brain regions essential for advanced intelligence. Previc writes: "The massive increase in dopamine during exhaustive physical stress in thermally challenging environments as well as the special role the dopamine-rich left hemisphere plays in lowering body temperature are also indications of an important role of dopamine in this regard."[8] Thus, the critical involvement of DA in brain regions facilitated human intelligence and played a crucial role in pushing the emergence of our cognitive capacities to higher levels of functioning. To this effect we must add that higher DA concentrations in neural reward structures bring about pleasant experiences that dampen arousals of fear and anxiety and so provide the emotional underpinning for the exploration of novelty and the courage to venture into new geographic regions, even over long distances.

The second factor was the decisive switch from a predominantly plant-based diet to a diet richer in protein (meat and seafood), which provided the nutritional underpinning for this development. The millions of seashells found near the Blombos Cave in South Africa bear silent testimony to the fact that our ancestors who lived by the sea made ample use of its bounty. Such a diet would have provided them with an increased intake of proteins, iodine, and fatty acids, necessary for brain growth and for DA production.[9] These mutually interactive factors help us understand the emergence of human higher intelligence as a relatively recent epigenetic response to environmental stresses (less than 100,000 years ago).[10]

As important as this biological argument is, it alone does not offer a satisfactory response to questions posed by our insatiable thirst for understanding and meaning: "Who am I?," "What am I here for?," and so on. I approach this issue by returning in the next section to the example of the prehistoric hunter who had marked the animal's jawbone with two oblique crosses.

---

8. Previc, "Dopamine and Human Intelligence," 321.
9. Previc, "Dopamine and Human Intelligence," 328n17.
10. Previc, *Dopaminergic Mind*.

Consciousness of Place and Power

From the ubiquity of the surrounding animal world, the hunter had chosen just this one animal, took aim with his spear, and killed it. Straining with all his physical and mental energy even at the risk of his life, the hunter had brought it down. The place where it died, where it now lies, and where the hunter is about to imbibe its flesh and with it its fascinating powers, is for him the one place specially marked out from among countless other places in the vast compass of his horizon. This place is now etched in his memory, defining a unique reference point, a point of orientation within a world of almost infinite possibilities. This hunter was thus answering the primary question of existential space: "I am where I act." It is the place where I, after mobilizing all my powers and abilities, meet and conquer a challenge posed by my existential needs in my environment.[11]

The German theologian Eugen Biser echoes the same thought when he maintains that the primary issue for humanity is the question of place—not geographically but existentially, and thus biblically—and that this question is also the question with which God addresses us.[12] In Genesis 3:8–9, the paradisal pair, after their transgression, were hiding from God among the trees of the garden. *"But the Lord God called to the man and said to him, 'Where are you?'"* Thus, the primary existential question is not whether and where I exist as a mere organism, blindly obeying biological urges, but whether I have found the place where I am enabled to lift my head, look and reflect on the horizon of my existence, and be sustained in the presence of the Creator.

The prehistoric hunter marked the place where he had acted by putting his signature on the kill with two oblique crosses on the jawbone: here is my place where I respond to the existential demands I must meet in my environment; the place of my action is where I experience a sense of being and meaning that provide stability and orientation when more distant and intangible horizons beckon. It is the place where through my free action I carve traces into the reality of my world and so create the place in which I am at home, and through my ever-new action I can create such a new place again and again, just as I am shifting my geographic horizons in the process. While these thoughts may be understood as touching on the existential significance of work in human existence, in

---

11. Baudler, *Das Kreuz*, 35.
12. Biser, *Der Mensch*, 37.

the prehistoric context they relate primarily to a growing sense of emancipation, the move from powerless scavenger to powerful hunter.[13] With growing cognitive faculties, our forefathers left the relatively food-rich niche of gatherers by converting their digging sticks into spears with fire-hardened points and turned to big game hunting. With time, they outdid even the largest and most powerful predators. They slaughtered the mighty mammoth, rhino, giraffe, bison, wild horse, giant elk, ibex, lions, panthers, and bears and so catapulted themselves to the apex of the food chain. Early humans had become *Homo necans* ("man the killer").[14]

It was this place in the total scheme of things that gave them a sense of being. As they sought more of it, they concentrated all their energies on bringing down the largest creatures. Using their growing working memory and their higher mental faculties, they would have imagined and literally prelived the excitement of the hunt with its dangers and risks, as well as the act of killing, which culminated in the spilling of the warm-running blood. Killing the animal was an almost cultic event, according to cave paintings of that period. In their cave paintings, these Paleolithic hunters often marked the images of animal bodies with rows of oblique crosses, usually on their most vulnerable body parts, the soft flanks and bellies to which their spears and arrows were aimed. Paleoanthropologists view these paintings not as collections of hunting trophies but as cultic symbols: the slain animals served as special food that, when imbibed in cultic-sacrificial rituals, would transfer the victim's powers of being to the eater.

It would be a grave mistake to assume that our ancient forebears possessed a modern consciousness that is able to see individual entities, objects, or beings as one of its kind among many, even if within the greater totality of nature. Rather, they perceived all-of-life, all-of-being, as one. The individual animal they killed and ritually consumed was not grasped as a single body among many other such bodies. Neither was nature conceived in abstract categories by regarding it as a mere food source. Instead, they would most likely have perceived these realities with an extraordinary intensity as something that transcended their tangible world. In moments of such esoteric experiences, they might have felt a kind of creaturely ecstasy that raised their consciousness from being merely embedded in a tangible world to one of an all-encompassing whole.

13. Baudler, *Das Kreuz*, 36.
14. See Burkert, *Homo Necans*.

Put differently, when face to face with large animals, our early ancestors would have encountered the same formidable macho power of nature that evokes exaggerated me-too power displays from male chimpanzees—chest pounding, stomping the ground, grimacing—during severe thunderstorms or blazing forest fires. Through the killing and consumption of big game, our prehistoric forefathers sought to absorb this fascinating power into their existence, believing it to be of supernatural origin. With this power at their disposal, the hunters themselves felt they had acquired the very qualities they revered, "horns like a wild ox" (Num 23:22).[15]

The attentive reader would have had an inkling that the foregoing descriptions, explanations, and reflections make use to a large degree of so-called inner experiences, an element that modern materialist science, with its commitment to the myth of pure objectivity, vehemently denies. Yet human cognition, as it turns out, is based on the experiences of personal subjects; it cannot be understood without it. What is more, subjectivity is not limited to human cognition. Not only are individual animals conscious when they experience satisfaction or suffer deprivation, but, as John Haught has pointed out, a "strand of cosmic insideness has gradually thickened and intensified since the origin of life less than four billion years ago and a silent anticipation has been part of the eventual emergence of the sentient and conscious interiority has been part of cosmic process from the very beginning."[16]

This "insideness" is the focus of all that follows in this chapter, beginning with its neuronal substrate, the human brain. As we will see, the brain's complexity, plasticity, and holistic functioning influence the way we perceive the world around us—including the transcendent dimension of the world—in mysterious ways.

The Human Brain—Complexity and Mystery

The main organ in the human nervous system is the brain. Protected by the skull, its largest component is the cerebrum, divided into two hemispheres. Under the cerebrum is the brainstem, behind which is the

---

15. We should therefore not be surprised to find that the primary religious symbols of storm and weather deities during the Paleolithic were the bull and the horse. Even Israel's deity was initially known as the bull, *El*, and worshipped in ancient Palestine at two locations, Dan and Bethel.

16. Haught, *New Cosmic Story*, 69.

cerebellum. A thick layer of brain tissue or the cerebral cortex stretches across the cerebrum. For comparison, taking body size into account, the human brain is almost twice as large as that of a dolphin and three times as large as that of a chimpanzee. A critical region of the human brain is the frontal lobe of the cerebral cortex, which is utilized in executive functions such as planning, reasoning, abstract thought, and self-control. The visual lobe, responsible for processing visual sense data, is greatly enlarged in the human brain compared to other mammals. The cerebral cortex consists of neural tissue folded into ridges (gyrus) and grooves (sulcus), making it possible to fold such a large amount of tissue into the available skull volume. In other animals, brain tissue is smooth.

The human brain weighs about 1.3—1.4 kg, making up 2 percent of body weight. However, for its size the brain is the single largest energy consumer in the human body. It uses 20 percent of the total budget, two-thirds to keep neurons firing and sending signals to other parts of the brain and to the body, and the remainder for housekeeping or cell maintenance.[17] Cell counts have shown that an adult brain contains 86 billion neurons, with an equal number of nonneuronal cells. Overall, the human brain is understood as a self-organizing system that generates the dynamic interaction of large-scale neuronal networks, mainly in the cerebral cortex and other circuits that facilitate human perception and cognition. Signal transmission, using biochemical and bioelectrical processes, takes place between nerve cells by direct connections, often over long distances.

That we humans possess extraordinary computational powers is widely recognized, although the complexity of these powers mostly escapes us. According to brain network analysts, at any given time the brain must solve two tasks simultaneously and as efficiently as possible: *segregation* of a vast range of specialized external and internal signals, and their *reintegration* into meaningful wholes. Given the huge number of neuronal and synaptic links in the brain, the constraints involved in these tasks push the connectivity of the system toward extraordinary levels of complexity.[18] Its astronomical intricacy becomes even more amazing when we consider that the 86 billion neurons of our brain fire at a rate of 10 times per second every moment of our lives! Such awe-inspiring

---

17. Swaminathan, "Why Does the Human Brain Need So Much Power?," http://www.scientificamerican.com/article/why-does-the-brain-need-s/

18. Sporns, "Network Analysis," 56–60.

inner wiring is needed to keep you and me alive, to facilitate our effective learning, and to orient us in and help us adapt to the world around us.

Once we become conscious of our own marvelously crafted brain, we come away humbled by its mystery, deeply respectful of both its complex architecture and of what it takes for conscious and intelligent life to exist on the planet. When we add to this breathtaking picture the countless selective moves that had to occur since the birth of the universe, from the formation of galaxies and the nucleosynthesis inside stars to the first appearance of a living cell and on through all the stages of growth that formed the many branches of the tree of life, we are left gasping at the mystery of our existence. Who would have thought only a few decades ago that we owe the substance of our bodies to the stars? As we saw earlier, their burned-out remnants scattered 10 billion years ago into outer space to form the chemical base for life. Not only has the universe used stardust to form our brain tissue, but this bewilderingly complex lump of brain cells is alive and can produce conscious states that can even think about and reflect on these things.

Besides, the brain can generate exhilaratingly wonderful experiences in us from mere sense perceptions. Just think of the fragrance of freshly cut grass, the thrilling elegance of bird flight, a baby's first smile, or the strains of a Mozart piano concerto. Colored with emotions as they are, we can laugh and cry, talk and listen, and when falling in love even surrender our entire being to another. Without these countless states of consciousness, we would barely consider ourselves alive. Yet, while we can describe such inner experiences, no one knows how the brain turns sense data into such ecstatic states of mind. We are simply left with an inscrutable puzzle, the enigma of human consciousness. Somehow, all these sensations break into our awareness when the right quantity and quality of signals are generated in our brains, leaving us breathless yet also curious as to what may drive this intoxicating richness.

At one level, the cause lies in the brain's relentless hunger for *significant* input, as it is one of the brain's main tasks to make meaning and generate understanding. While we do not know how, we know that the brain does do these things. Take our use of language as an example. Logically, we cannot derive legal concepts from our speech about ordinary human actions. Yet we can and do regularly grasp legal meanings when we encounter them. How do we do it? We don't know, but we do it.

As a meaning-making organ, one of the brain's outstanding features is that it is not a passive recipient of mere stimuli. Rather, it is constantly engaged in the business of sensory perception and goes about it with a ravenous appetite that looks for *patterned information*. As neuroscientists tell us, the brain is never idle but is constantly busy looking for and making up patterns; it even asks itself questions. This active curiosity promotes a sharpened awareness, or what is known as *attention focus*, which amazingly energizes our consciousness.[19]

This remarkable feature is vividly on display in children as they explore the secrets of their world by relentlessly asking questions. In this way, they activate their consciousness and focus their learning. Hungrier for meaningful input than for facts, their brains generate questions that often reach way beyond mere factual answers. Even questions that on the surface look like queries about facts hide a child's deep longing for meaning: "What makes the flowers grow?," "Where does the weather come from?," "Why did the cat die?," "Who put the moon in the sky?," "Why do birds lay eggs?," "Where do I come from?," and eventually "Why am I here?" But more on that later, when we consider their spirituality.

Questions like these would have stirred our earliest ancestors. Like all who have pondered them since, they had to have reached for more than facts. They sought meaning beyond the factuality of physical phenomena. Why? Because our brains act as if we are somehow connected to something greater than ourselves, to a horizon where perhaps ultimate meaning may be found. Interestingly, this perspective is underscored by the findings of modern brain research and neuroscience. Before widening the exploration, let me first go deeper into the fascinating subject of the brain's function at the level of its neurons.

The Brain and Its Neurons

The question of how our brain generates conscious mental states is not new. In neurobiology and philosophy it is known as *the mind-body problem*. While several theories exist, mainline neuroscience has opted for a model that sees the brain and its functions in terms of classical causal behavior, at least for now. It holds that the brain functions like a system in classical physics based on cause and effect. Simply put, the brain generates noisy dynamic behaviors that then cause a variety of conscious experiences. Neuroscience thus assumes that conscious experiences have

---

19. Bor, *Ravenous Brain*, 75–77.

been caused by neural activity, and this assumption underlies the neuroscientific research agenda.

Cambridge University neuroscientist Daniel Bor offers a clear description of this so-called classical model: "*All conscious and unconscious mental processing equates to the electrical activity of vast collections of information-processing brain cells called neurons, each with a biological version of thousands of input/output wires connected to other neurons, thus allowing each neuron to be connected to other neurons thus allowing each neuron to influence, and be influenced by, the activity of many others.*"[20] Biologically, the body of each neuron cell is surrounded by a multitude of branches that function as input/output wires linking neurons to other neurons. Each neuron features only one input wire, but many output branches that transmit signals to multiple neurons at the same time. The signals consist of a simple electrical charge measurable in standard voltage. These charges obey a binary code; nonfiring means 0 and firing means 1. Remarkably, the ends of the wires do not touch the neurons. The gap between the neuron and the wire is called a synapse. When neurons fire—that is, send a signal—they release in the synaptic gap a puff of neurotransmitters, which are picked up by the other neuron.

Many types of neurotransmitters cruise in the brain, some designed to suppress action, others to enhance it. Depending on the chemicals released in the synaptic gap, certain numbers of neurons may fire together. Where this happens frequently, the strength of these connections increases, making it more likely for them to fire together in the future. This feature has important implications for focusing attention. We now know that information (and memory) are not held in specific brain locations as in electronic computing, but in patterns of neuronal activity, partly encoded in the strength of connections between neurons of huge networks that involve millions of information-processing brain cells.

Neuronal networks built from links with other neurons are key to the brain's way of learning. These networks may start small but can scale up exponentially. Bor writes: "This apparently simple neural system allows for incredible flexibility, especially in its larger forms. Just as DNA is written in a language understood by all life on earth, [the brain's] ubiquitous neuronal binary language is potentially understood by the whole brain, which, for instance, allows for the easy exchange of information

---

20. Bor, *Ravenous Brain*, 119, emphasis original.

between [brain] regions."[21] This highly malleable feature is quite consistent with and reflective of the brain's evolutionary origins. How deep this relationship goes was noted in studies by astrophysicist Franco Vazza (University of Bologna) and neuroscientist Alberto Feletti (University of Verona). Comparing the density distributions in the neuronal network of the human brain with simulated images of filament networks in galactic systems, they found describable similarities. Both systems are arranged in well-defined networks—nodes in the brain and galaxies in the universe—connected via filaments, albeit at vastly different scales, the one between one micrometer and 0.1 millimeter, the other at a scale of five to 500 million light-years.[22]

Because the brain filters out all but a small fraction of information from the vast sweep of stimuli that enters our senses, we are conscious of only a tiny subset of the world around us. This tight focus is advantageous in several ways. It allows us to spot noteworthy changes and sudden events in our immediate surroundings. Further sharpening our attention, it offers the opportunity to observe and deepen our knowledge of a limited number of features of our world, which in turn enables us to consider and perform specialized tasks in relation to these features.

What we are not aware of, however, is the aggressive filtering of the information coming from the environment, which constantly provokes our brain cells to engage in frenetic activity as vast networks of firing neurons compete for attention. When this battle is won by those neurons that shout loudest, their signals break through and reach our consciousness, shutting down all nonrelevant inputs at the same time.

Far from comprehensive, this brief sketch must suffice as a glimpse of the complexity and mystery of our brains, the cradle of our consciousness, as well as the recipient of revelation. With this latter assertion, we touch on a question implicit from the beginning of this chapter: Where do divine revelations fit into this neuroscientific description of human consciousness? I propose to approach this subject in three steps: first, by considering the brain's involvement in generating beliefs; second, by looking at the influence of the brain on religion; and third, by inquiring into the brain's role in forming conceptions of God.

---

21. Bor, *Ravenous Brain*, 122.

22. Starr, "Study Maps," https://www.sciencealert.com/wildly-fun-new-paper-compares-the-human-brain-to-the-structure-of-the-universe.

Brain and Belief

Beliefs are not limited to religion. As habits of mind, we find them involved in all aspects of life, especially whenever it is necessary to exercise trust or place confidence in a person or thing. Hence, our beliefs are reflected in all cultural and societal institutions like politics and economics, customs and traditions, morality and ethics, and, of course, in expressions of religion and spirituality. Our beliefs therefore suffuse every part of our lives and so largely shape how we vote, spend our money, and decide where to live, as well as determining our choice of occupation, partner or spouse, and possibly childrearing practices. That every one of these beliefs is a product of our brains and that without our brains none of them would exist we seldom consider.

But why does the brain create mental states we call beliefs, and what has brain research to say about the extent that we, as a species, may be naturally oriented toward a transcendent interpretation of our sense perceptions?

In evolutionary terms, beliefs exist because they help us effectively navigate the complex world around us. In short, the ability of the brain to form beliefs has survival value. When so engaged, the brain must first segregate or filter vast amounts of information from our sense perceptions for relevance. In this meaning-making effort, the brain works with a limited amount of data compared with the totality of information out there. To create a reliable picture of the whole from limited data whose trustworthiness is not always clear, the brain activates its belief-making ability. Like connecting the dots in an incomplete picture, the brain fills in the gaps, making up beliefs from stored information. We might say that the belief-making feature of the brain comes into its own whenever it must work in an underdetermined context or in the gap between things that can and cannot be proved.

While this idea may be relatively easy to grasp, understanding how the brain goes about making sense of the data reliably is far more difficult. Let us take it step by step. Defining "belief," we would say that it is a sense perception, cognition, emotion, memory, or, more generally, an awareness that a person accepts consciously or subconsciously as true. As a self-organizing system, the brain constantly constructs perceptions of reality from whatever data it has available—consciously and subconsciously. If it receives conflicting or insufficient data that does not agree with the stored picture of reality, it will fill in the gaps by forming tentative models or beliefs about them. As soon as the logic of the perception makes sense, our cognition goes about seeking proof for the brain's working hypothesis. In this effort the brain relies on one of its most powerful tools for creating beliefs, our memory. And to maintain beliefs the brain builds up memory banks by repetition. For instance, if our memory has registered a hurtful encounter with another person, it will serve as proof for the belief that we are dealing with a hostile customer who is better avoided. The same goes for experiences of positive emotional content.

Because our memory is not always reliable, this process of belief building is not error free. Let me illustrate what I mean with an example. The visual cortex participates significantly in the belief-generating processes of the brain. When the brain deals with visual patterns that are difficult to decipher, it will fill the gaps in the cognitive picture based on past visual imprints. Depending on what is stored there, our visual perception may suggest that we have seen curved lines although they were in fact straight, which is the phenomenon behind optical illusions. Memories can be very powerful and may override other inputs, especially when strong emotions are attached to them, for in the language of the brain the importance of information is signaled by the weight of its emotional content, so that strong emotions tend to elicit strong beliefs. The stronger the emotions, the more important the information and the stronger our defense when the beliefs are challenged. Conversely, perceptions and thoughts, if their emotional weight is light, may never reach the level of consciousness in the first place.

Yet, strongly held beliefs based on powerful memories may not necessarily be accurate. Since the brain learns only by trial and error, it cannot a priori distinguish between what is true and what is false. What protects us against life-threatening random errors are so-called social safeguards, meaning two qualities with deep roots in our preconscious past. On the one hand, we don't exist in isolation; on the other, we are highly imitative creatures (conditioned too by the geography and landscape in which we were raised). As a result, our memories and our beliefs are strongly influenced by others with whom we interact throughout our lives—parents, siblings, peers, social networks, church leaders. In addition, we are all subject to the collective mind of society and its culture. This group consensus shapes our perceptions of what is true or false, what is or is not desirable, not only for our survival but also for a measure of human flourishing. But, as history and life experiences teach us, even the collective mind of a society, culture, or religion is no guarantee that its beliefs are true, just, or good. For example, the collective mind may be deceived about the truth value of an ideology and suffer the consequences.

To better understand how our beliefs form and how they work in the brain given the multiplicity of perceptions, emotions, memories, and so on that we experience, let us return to what we have said about neuronal connections—that every time a neuronal connection is used it is strengthened, and when it is not used it is weakened. The same idea pertains to every thought we focus on and that supports a belief. Since the simultaneous firing of neurons leads to increased connections between the neurons involved, the more we activate the circuitry of a belief through repetition, the more we strengthen it. Conversely, when we reject behaviors and thoughts that contradict what we believe, the associated neuronal connections are weakened. It is this plasticity of the brain that creates the possibility of changing unwanted or harmful beliefs and affirming valuable ones. In the Judeo-Christian tradition, such a change of mind is known as "repentance," meaning in its ultimate sense returning to God in faith and obedience. The same plasticity is at work in building, shaping, maintaining, and dismantling all beliefs, whether religious, social, scientific, moral, or ethical.

To sum up, then, beliefs are constructed realities, derived from perceptions, emotions, memories, thoughts, and words in their significance for our lives. They are in part a function of the strength of neuronal connections that underlie the creation of these beliefs. At the same time, it is sensible to heed Andrew Newberg's caution: "While the brain does a great job of formulating our beliefs that work well for us and help us understand the world, beliefs are also fraught with many potential flaws. The brain can [and does] make mistakes at many different levels while constructing its beliefs."[23] The message from neuroscientists is simple: consciousness and its beliefs depend on brain functions, conscious or unconscious, including all the religious notions we can experience. Religious ideas, and therefore revelations, are structured as well as limited by what an individual brain can do within the cultural context in which it was nurtured. And, as neurophysiology and neuropsychology would assert, this is valid for the definitions and criteria we use in describing religious ideas and spirituality as well as perceptions of "the sacred," "the divine," "ultimate reality," or "ultimate truth."[24]

---

23. Newberg, *Spiritual Brain*, 116–17.
24. Newberg, *Spiritual Brain*, 118.

Given the structure and functioning of our brain, it is not surprising that religious pluralism should arise in the world. This prompts us to ask how religious and theological ideas emerge from certain brain processes.

The Brain's Influence on Religion

With the foregoing in mind, it stands to reason that brain processes do bear on religious and theological concepts as they bear on everything else we conceive. Keeping open the possibility that a theologian or sociologist might choose different definitions, there are basic ways the brain processes data that pertain to religious and theological concepts. To illustrate, brain researcher Andrew Newberg takes us into a brain region—the inferior parietal lobe—an association area of the brain located at the junction between the visual, auditory, and body-sensory pathways—which, among other complex functions, enables us to discern cause and effect. Newberg writes: "When this causal function is applied to the physical world, the result is the scientific method. When [it] is applied to the human world, the result is social science. When the causal function is applied to the spiritual world, the results are concepts related to God or ultimate reality."[25]

What if we did not possess this causal function? While we would be able to think of God in terms of some of his attributes—as holy and loving, for instance—we would be unable to understand God as the cause of all things. Here we must remember that what our brain does has no bearing on reality as it is or on the question of whether reality conforms to what the brain perceives. Whether the brain "believes" in gravity or not does not affect its force in the universe. In the same way, brain functions do not provide proofs of the existence or nonexistence of God.

The brain region responsible for rational thought and the ability to form abstract categories and symbols is the superior temporal lobe. This function, when applied to the religious or spiritual realm, allows us to consider religious and theological concepts and discuss religious practices and their meaning, as well as matters involving morality and ethics. Also, when analyzing concepts and problems, it is this abstract function that comes into play. When the causal and the abstract functions work together in binary mode, they help us create pairs of opposites like good and evil, right and wrong, God and the world. Newberg also sees in the

---

25. Newberg, *Spiritual Brain*, 119.

latter the fundamental binary structure of God versus man at the root of all religions. It is the aim of religion to reconcile this binary structure; to accomplish it, religions follow a holistic path (involving faith, hope, love, prayer, meditation, forgiveness, and so on). This is supported by the brain's holistic capacity, a function that takes place in the right hemisphere of the brain. When activated during meditation and prayer, for example, it blocks out certain sensory inputs, leaving behind an experience of reality as being whole. Since the sense of oneness may vary, Newberg alludes to the possibility that there may be a "unitary continuum of human experience and thought, including the emergence of the sense of oneness as a property of smaller processes, a oneness of God, and a oneness of all things."[26] In other words, the concept of God as unifying force has deep roots not only in religious history, but also in the interplay of brain processes that mediate a sense of oneness, and, let me suggest, in the fundamental structure of the universe itself.

Finally, we must consider the brain function that opposes wholeness; this is the left-brain reductionist function. Located in the temporoparietal junction, it helps us reduce things to their constituent parts. Applied to the religious realm, it takes a fundamental story or founding myth and breaks it down into many parts and attempts their interpretation, including through liturgical and other forms of religious practice. In short, the causal, rational, binary, abstract, holistic, and analytical functions of the human brain all have their place in helping us make sense of our external world and of the world of our inner experiences.

We should mention one more cognitive system crucial for the ordering of our thoughts, especially when it comes to formulating religious ideas: the emotional system. As noted earlier, it is the emotional weight of our experiences that preserves them in memory and consciousness. This prioritizing function helps us establish a hierarchical ordering of our religious and nonreligious thoughts and concepts. Theologically speaking, the emotional system is related to experiences we call love, compassion, joy, and peace. Identical brain processes are involved in the Judeo-Christian concept of redemption, which includes the binary notion of salvation, of being saved or not being saved. According to Newberg, it is the emotional component attached to the experience that renders these notions so potent and durable.[27]

26. Newberg, *Spiritual Brain*, 120.
27. Newberg, *Spiritual Brain*, 122.

As we surveyed various brain functions and their application to the religious and spiritual realm, we saw how their interplay promotes the formation of such notions as ultimate reality, God, and the relationship between God and us humans. Since our brain is foremost a meaning-making organ, it will always, irrespective of the task to which it is put, seek to construct a functionally whole reality for us. With these thoughts in mind, we turn now to the more comprehensive question of how our various cognitive and emotional processes affect our beliefs in God.

The Brain's Influence on Conceptions of God

In 2006, a research team from the Institute for Studies of Religion at Baylor University in Texas published the findings of a major study on religion in the United States.[28] They surveyed more than 1,700 people, each answering 400 questions. The results highlighted the complex ways people think about God and spirituality, whether they believe in a divine dimension or not.

American beliefs about God seem to fall into five primary categories: that God is authoritarian (31.4 percent), critical (16 percent), distant (24.4 percent), benevolent (23 percent), and nonexistent (5.2 percent). According to the study, these perceptions are affected by several sociocultural factors: gender, education, income, and geography. Women tend toward engaged images of God (nonauthoritarian and nondistant), while men tend see God more as distant or as nonexistent. Those with lower education and income tend toward a more engaged God, while for those with higher education and incomes above $100,000 God is more distant or nonexistent.

The study also reports on the influence of geographic demographics on how Americans see God: Easterners tend to see God as critical, Southerners tend toward an authoritarian view of God, Midwesterners tend to see God as benevolent, while West Coasters tend toward images of a distant God. In short, when it comes to questions of what God is like, people offer a variety of answers that seem influenced by a variety of physiological and social factors.

Seen from a neuroscientific perspective, these results are not surprising. After all, the brain receives millions of pieces of information and

---

28. Bader et al., *American Piety*, 26–30, https://www.baylor.edu/content/services/document.php/33304.pdf.

seeks to create a coherent version of what our world is like. Integrating what we see, hear, taste, and smell, the brain constructs a picture of the world around us, *re-presenting* to us its own interpretation of the data it receives. Here again, we must remember that the brain's constructs do not necessarily correspond to how the world really is. The same is true for the brain's perception of spiritual reality and God.[29] Newberg writes: "Whether or not there is a God, and regardless of what attributes God might have, each of our brains interprets this information for us, and our interpretation is constrained by the structure and function of the brain. In other words, we make of God what our brain allows us to make of God."[30] To the degree, therefore, that our brain is unique, even if we share a common ethnicity, geography, and culture and so hold similar perspectives on life, our interpretation of reality, including God, will also be unique, as the Baylor survey shows. I explore the significance of this conclusion for the concept of revelation below.

For now, we want to ask how the brain goes about the task of thinking about what God is like and whether neuroscience has anything to say about the biological underpinnings of religious conduct.

As we have seen, a neuroscientific approach to the God question proceeds from the cognitive functions of the brain. Just as these functions help us perceive other aspects of the world, they help us understand basic religious ideas and behaviors as well as specific ideas that pertain to God. Besides, we have already noted the significance of Newberg's functional designations: holistic, reductive, causal, abstractive, and binary. To this we must add the all-too-human tendency to conceive of God in anthropomorphic terms (as many people do regardless of their denominational affiliation), imagining God as a superperson, often as an old man above the clouds. If people visualize God as a person, neuroscience argues, it is a habit primarily supported by the predominance of the visual cortex. In a similar vein, the tendency to assign a human face to God is readily explained by the brain's capacity for perceiving and interpreting facial patterns.

We not only tend to think about God in visual terms and give God a human face, but we also communicate with God as we would with another person, in conversational prayer. The latter activates the brain's

---

29. When misinterpretations by the brain generate misbeliefs, these—if acted on—will usually lead to painful life experiences that work like a feedback mechanism, checking the brain's reliability and helping us to correct shortfalls.

30. Newberg, *Spiritual Brain*, 140.

speech areas, which, if coupled with positive emotions, for example, would likely lead to perceptions of God in positive emotional terms, a point the Baylor study also confirmed. In short, the brain believes what it perceives is real, and we should not be surprised if we tend to assign reality to the extent that something seems real to us. On the other hand, it is not clear to neuroscience why we think that anything should or shouldn't exist, although the brain's binary operator would theoretically be able to distinguish existence from nonexistence.[31]

Of course, none of the above observations can be used to justify a specific religious belief or to support claims about the existence of God. The one element that, more than any other, has been used to warrant such claims, however, is the phenomenon of mystical experiences. Before we explore what neuroscience has to say on this subject, let us consider a phenomenon rarely considered when working our way toward understanding religious or spiritual matters in terms of science and revelation: the spirituality of children.

The Spirituality of Children

In the 1990s, David Hay and coworkers dialogued with a group of British children who were between six and 10 years old about spirituality.[32] After eliminating cultural-religious biases from the children's responses, they analyzed the children's awareness with the help of three neutral indicators of spirituality—awareness of the Here-and-Now; awareness of Mystery; and awareness of Value—grounded in insights from developmental psychology that, in comparison to adults, babies and young children are relatively more conscious of their external environment than of their inner life.[33] Also, across all traditions, masters of the spiritual life have emphasized the practice of *presence* or the mode of mind that is occupied neither with the past nor with the future, but with the now. When adults attempt to reclaim this mode of being they had vividly experienced in infancy, Hay suggests, they engage in the spiritual practices of contemplative

---

31. Newberg, *Spiritual Brain*, 148.

32. David Hay (1935–2014) was the director of the Religious Experience Research Unit at Oxford University (now known as the Religious Experience Research Centre at St David's College, Lampeter, University of Wales) and honorary senior research fellow at the University of Aberdeen. His publications include many books and articles. Of special interest here is his *Something There*.

33. Margaret Donaldson, *Children's Minds*.

prayer and meditation.[34] Moreover, in most cultures children are considered deficient adults, whose ignorance of how the world works must be corrected through education. In modernity, the aim is to provide them with intellectually satisfying explanations for every conceivable phenomenon. The Western educational model especially seeks to ensure that by the time they leave high school students have been firmly inducted into a world devoid of mystery. In this new and disenchanted world, everything is described by proffering a scientific explanation. Yet a cultural paradigm that sees a child's mind as deficient and technical knowledge as the only knowledge that exists is open to challenge. German philosopher Martin Heidegger has rightly criticized Western thinkers for forgetting "Being," arguing that they have consigned the entire philosophical tradition to a state of amnesia, as it no longer has an answer for the typical question a child asks: "Why is there something, where there could be nothing?"[35] Awareness of this cultural neglect, and of the relentless curiosity of children that points to a spiritual dimension, persuaded Hay to include "mystery" as a research category.

After analyzing more than 1,000 pages of children's interview transcripts, Hay and his colleagues reported their amazement over what they discovered. Quantitatively, except for one or two cases, all the children reported the presence of a spiritual dimension. Qualitatively, the data "engendered a powerful cumulative sense of being presented with a common universe of human experience [of spirituality]."[36] Line-by-line analysis of the transcripts had identified an unusually high level of perceptiveness in the children's responses that pervaded the entire context of how they related to their world, leading Hay to call it by a special term, "relational consciousness."[37] As early as the 1960s, famous Oxford zoologist Sir Alister Hardy proposed that in human spiritual awareness we encounter a universal, possibly primordial competence. While Hay's results are very much in line with Hardy's hypothesis, the latter may also be taken as a contemporary echo of the biblical intuition that humankind was created in the image and likeness of God (Gen 1:26–27).

---

34. Similar observations come from Richard Rohr, director of the Center for Action and Contemplation in Albuquerque, New Mexico, who calls the contemplative mind the "beginner's mind" (Rohr, *Everything Belongs*).

35. Hay, *Something There*, 132.

36. Hay, *Something There*, 138.

37. Hay, *Something There*, 139.

The conclusion? This characteristic of relational consciousness is grounded in our biology. We naturally possess this unusual relational capacity, because of our biological constitution and our processes of becoming, so that we are spiritual on physiological grounds. What does this mean? If we happen to be theists, this relatedness allows us to be aware of God; if not, we are likely to perceive "a seamless holistic relation to the Other, whether conceived in secular or religious terms."[38] This conclusion draws attention to what some scientists have long suspected. It is incoherent to call material substances capable of generating subjective experiences like relational consciousness" "mindless." The legitimacy of this objection shall occupy us again in the final chapter.

Since this holistic perception that everything is related to everything else—that "everything belongs"—is also the mark of the infant's mode of being, we begin to see why this state of mind belongs universally to all humanity. All of us were born with it, and we all once lived in it. We all were once held for nine months in a tiny sac in our mother's uterus during her pregnancy. Bathed in the warmth of amniotic fluids, we were totally one with her, from conception embedded in an unimaginable intimacy, a bond that even the birth process does not sever. And, as all mothers know, babies actively participate in this all-encompassing relatedness, reciprocating soon after birth by imitating the emotional signals their mothers lavish on them. Later, out of these physiological processes our alert openness to the world develops, to which our primordial state of being contributed a deeply *relational consciousness* at the base of our spirituality.[39]

The Mystical Mind

Among the wide range of human experiences, mystical experiences rank among the most vivid and intense, as modern brain studies have shown. People undergoing such experiences report that the distinctions between self and the other, between inside and outside, had completely disappeared, and they experienced oneness instead of barriers, a unity of all things, even at a cosmic level. Moreover, this perception of oneness tends to be accompanied by a sense of great tranquility and peace, evoking calming effects on the autonomous nervous system.

38. Hay, *Something There*, 140.
39. Hay, *Something There*, 141.

When asked how neuroscience approaches an experience that perceives an absolute unity of all things, or how the brain may be able to conceptualize a grand notion such as the oneness of the cosmos, Newberg points to another experience that involves this universalizing capacity of the brain—the experience of love. Love, erroneously understood merely as an emotion, involves several brain areas, including the reward system and social-relational perception. Apart from helping us perceive our emotions, these two brain capacities enhance our ability to connect with others. Because love is after all about connectedness, about unity and wholeness, Newberg infers that the brain may not just be an organ capable of spiritual experiences, but that it may be mystical before all else. The mystical mind, understood as an appetite that ravenously seeks to achieve panoramic coherence, works incessantly to integrate biological, psychological, social, and spiritual elements. This intense work of integration that melds inner experiences with data from the external world involves billions of neurons providing us with a sense of a unified view of reality and mediating to us a sense of identity and orientation in the world. In short, it determines who we are. The integration may be so complete that we are barely aware of boundaries—for instance, where the brain ends and where the world begins.[40] Summarizing his findings about the spirituality of children, Hay spoke in similarly panoramic terms of an unusual level of perceptiveness that encompassed the entire context of how the children related to their world.

Hay's insights, especially the latter, suggest the presence of a certain mindset that enables the experience of a fundamental sense of belonging to a mystery greater than ourselves that we cannot control. In Jesus' teaching, one of his favorite visual aids is a child. When his disciples got into power games and rivalry, he confronts them with the lowly position of a child (Matt 18:2–5). In Jesus' language the child symbolizes the original or primal knowing, the unitive experience of an infant, even from before its birth, with its mother. In the womb, the baby knew only one reality: infant-mother oneness.[41] Religiously, this experience is described as "knowing in faith." Mystics know it as "unitive consciousness," a kind of panoramic awareness and fundamental openness that sees the whole world as connected and ourselves as part of it. In Saint Paul's terminology,

---

40. Newberg, *Spiritual Brain*, 158–59.

41. Consciousness of the baby's separation belongs to a later stage of the child's development.

it is the "mind of Christ" or "the mind that receives all things." Poet and nature mystic Anne Hillman writes about such an experience:

> Life tugs at our hearts like a magnet. It allures us—draws us into relationship with all that is. It was the *allurement*—this power of attraction calling everything into communion—that seduced and sang to my whole being. I was attracted to life the way the earth is attracted to the sun, the tides to the moon, and flowers to light. From some place deep down in my body, I knew from this yearning in my bones that everything in existence was utterly woven together.[42]

This mind has been at work in mystical experiences during all ages and cultural epochs. More significantly, perhaps, the brain itself is a oneness, composed as it is of billions of neurons that work together so that a holistic consciousness emerges, providing us with this unified view of reality. Put differently, this state of human consciousness, including its content of a unifying subjective experience, has been marvelously generated by the biological functioning of mere brain tissue. We have no way of describing how this happens, although we may, in utter astonishment, advance the thought that this ineffable indescribability of mystical experiences analogously befits the physical structure of the brain itself.

Here is one more step I want to take in this exploration of the enigma we call the mystical brain. Because the brain's causal function thinks that the oneness it experiences is itself the cause for the material universe, a theist may be inclined to assume that God is the creator of the universe. The same person may also experience a sense of otherness as the brain's binary function differentiates between mystical reality and the tangible actuality of the here and now while the brain's emotional area mediates the perception of awe and joy.[43] Such an experience may be quite willingly and joyfully believed to be the experiential reality of God's presence, especially by those with a mystical leaning, in whose understanding God is not an objective fact or an actual being among other beings, but the absolute, undifferentiated Oneness, the ground of all being and existence. This experience opens up a huge cognitive gap: God is considered indescribable, yet the experience I have described deeply connects the individual with his or her notion of God.[44]

---

42. Hillman, *Awakening the Energies of Love*, 189.
43. Newberg, *Spiritual Brain*, 161.
44. Newberg, *Spiritual Brain*, 161.

Classical theism sees God as personal, knowable, and totally distinct from creation, and assumes that he makes use of standard brain functions and their cognitive capacities—including thoughts, speech, emotions, and behavior—to reveal himself. Mystics, by contrast, feel drawn into an intimate union with God, culminating in the notion that God is not elevated above us but is within us. Their experiences are extremely real and of such intensity that the individual is transformed. They experience not only deep changes in all aspects of life, in how they think, feel, and relate to the world, but also undergo profound and lasting shifts in consciousness, from separateness to deep connectedness, from self-centeredness to love.

Such panoramic, all-inclusive experiences of "big mind" have not always been welcomed by institutional religions. Those who center on rituals, doctrines, and dogmas are more at home with a conception of God who is distant and critical, hence their tendency to favor a mindset that analyzes, clarifies, and explains. The mystical tradition refers to such a mindset as "small mind," a descriptive term that is not meant to be derogatory but necessary, especially when clarifying theological meanings. Yet, to provide context and perspective, organized religion needs "big mind," for without the experience of a deep connectedness to all things organized religion loses the sense of mystery and the ability to gaze in wonder on the whole.

As we have seen, consciousness and mystical experiences cannot be denied, and both are intimately tied to the biological functioning of the human brain. Yet, when asked how this happens, we are at a loss to explain it. All we can do, in closing, is admit in humility that it seems beyond our ability to give a scientific account of the ineffable in mystical experiences, just as it is impossible to explain how mere brain tissue can produce mystical-spiritual consciousness.

Consciousness as Inside Story of the Universe

In several places I have touched on the question of how brain tissue can produce reflective and symbolic thinking, this distinctive human capacity. This question—known as the mind-body problem—has puzzled philosophers for centuries. Some of us, in a reflective mood, may have asked what it might have been like for our early ancestors to have experienced their first-ever thought, or the conception of the first-ever idea. We will never know, of course. On the other hand, it is the brain's primary function to interpret sense data to ensure survival. Even those with severe brain injuries have experienced the fruit of this mysterious capacity, enabling us to interpret and even attribute meaning to the world around us.

In our evolutionary universe, this ability did not appear all at once but emerged gradually over long periods from previous stages of development, in this case from protohuman brains. In this momentous instant in cosmic history, when *Homo sapiens* arrived, the very stuff of the universe was raised to a fuller level of existence. Afterward, although physics and chemistry continued to work as before, nothing would ever be the same in the entire creation. What had come forth was not a ready-made creature, but the fruit of thousands of evolutionary steps that have occurred over billions of years since the genesis of the cosmos, involving the formation of galactic superclusters, galaxies, and the tiny solar systems that gave rise to the biogenesis on which the emergence of *Homo sapiens* depended. Built-in creativity at the cosmic dimension ensured that entities of greater and greater complexity came forth to replace outdated ones that died out.

Once the level of biological organization was reached, even higher orders of complexity were achieved. Organisms with rising intelligence and consciousness appeared that even possessed the capacity to act intentionally toward future states. Hence, to understand human beings in evolutionary terms, we must see ourselves not as the exception in the process of evolution but as its recapitulation. Our capacities for thought together with our insatiable quest for meaning are thus understood as the gradual awakening of the materiality of the cosmos to reflective and relational consciousness. Let us consider some historical grounds for such an outrageous claim.

During the Lower Paleolithic, in the Acheulean Period (in Europe about 500,000 years ago), carefully hewn human stone artifacts had been produced that fitted snugly into a hand, so-called spheroids that were not tools or other objects of daily use. Creating a spherical stone object by shaping it with other, harder stones must have been extremely laborious. Since these objects seem to have had no other purpose than to be held by early human hands, we cannot escape the question of their symbolic meaning. The German prehistorian Marie König interpreted them as the oldest works of art by human hands. The artificers may have felt an inkling of something whole in the rounded stone, even perceiving it as a symbol of perfection and so by implication as a symbol of the universe.[45] Perhaps, Georg Baudler muses, sphericity may have resonated with these early humans, whose interior was awakening under the influence of ever-sharper perceptions of their environment. They may have seen the sphere as representing the distant horizon that encompassed them, including the dome-shaped canopy above, where clouds cruised silently by day and stars by night.[46] König sees the first inklings of a mental-spiritual culture here.

In this context, the work of the American philosopher Susanne Langer also comes to mind. Langer's study of human consciousness led her to conclude that humans possess an innate and pervasive need to invest their environment with meaning, and thus a capacity for symbolizing.[47] She argues that the awakening of the human spirit has its origin in the perception of symbols, and further that the emergence of this faculty may already be observed in the transition from higher primates to humans. Langer refers specifically to intelligence studies of chimpanzees. To illustrate, I am singling out a most impressive example: an adult female chimpanzee named Tschego, who cherished a naturally rounded pebble like a talisman. Under no circumstances would she part with this valued object, carrying it in a furry fold near her hip by day; at night she took it into her nest.

This behavior seems to suggest two things: the very first artifacts of this kind may have been found by our ancestors in nature, and possibly the wider spherical dimensions of the environment—the horizon, the sky above—may have captivated the awakening consciousness, imagination, and dreams of our early human ancestors. In the sheer sphericity of many natural objects, they would have noted, perhaps ever so faintly, that they were touching a horizon of wonder that lay beyond their needs for survival. This faint perception might have moved them to reach for it with their hands, sensing in its roundness the symbolic representation of a deeper, more awesome reality: the unity of all existence. As they reached beyond the limit of old perceptions, our early ancestors might have felt confronted with—as well as encompassed by—something grander, something intangible, elusive, even infinite. This dimension of existence, according to Baudler, gave rise to every religious-cultural and

---

45. König, *Anfang der Kultur*, 32.
46. Baudler, *Das Kreuz*, 28.
47. Langer, *Philosophy in a New Key*.

spiritual expression ever birthed in human hearts during our evolutionary journey.[48]

Yet our early ancestors' first encounter with this other dimension must have been frighteningly disorienting. Face to face with this intangible, even transcendent dimension, they would have experienced their existential precariousness at another level. Perhaps it was in such moments of vulnerability that they reached for a rounded pebble and that its roundness, as a symbol of the world's wholeness, transformed their aporia into a renewed sense of safety. Perhaps thus reassured, they returned from this liminal experience with an expanded vision. Certainly, the elusive boundary they had touched had opened out into a hazier, more ambiguous but also more promising reality. And from now on, every rounded contour they chanced on in their environment was invested with new meaning, just as every forward urge toward new horizons would be accompanied by a childlike reassurance that old limits are imaginary, that territories beyond them are inhabitable, albeit not without precariousness.

Ever since, humans have lived as much with symbols and meanings as with signals from their senses, in a world compounded by tangible things and virtual images somewhere between fact and fiction. Suspended between the impulses and interests of their mammalian nature—eating, sleeping, mating, seeking comfort and safety, avoiding pain and premature death, becoming sick and dying—our early ancestors now lived through symbols and perceptions of the transcendent that would lead them much, much later to laws and religion, to science, philosophy, and culture.

Now equipped with higher cognition and with a working memory, humans can hold their ideas in mind for reflection and expansion long after the occasions leading to them have slipped away. By associating symbols, mental images, signs, and words, humans can even think of things or plan and create future events not present in their actual environment. What is more, humans can shape their ideas, symbolize parts of some and discard others, even uncouple what they conceived from the original base through the power of abstraction. The combined effect of these faculties, which we call reason, is immense. Yet there is a downside. To be truthful to the human condition, we are obliged to explore this downside (in chapter 8) because human activity on the planet shows that

---

48. Baudler, *Das Kreuz*, 30–34.

there is far more unreason among *Homo sapiens sapiens*, the self-styled doubly wise human, than there is among our animal cousins.

Linking our species to prehistory has had a necessary educational purpose. On the one hand, it has helped us see human evolution in context at a time when many deny its reality. On the other hand, it has opened the door to another perspective that is generally ignored: with the rise of human consciousness an enormously significant, even dramatic element has appeared, because from now on personal subjectivity resides in the universe as the inside story. John Haught puts it more eloquently: "A vein of interior awakening has long been throbbing in cosmic history, but conventional science is not equipped to take its pulse."[49] This inside story, which cannot be measured with the tools of science, is told in the whispers of personal perceptions, subjective cognitions, desires, anticipations, and hopes—that is, in the stuff of experiences recorded by personal subjects. This insideness would be unthinkable without the outside story of the universe, its evolution over billions of years, the formation of its macro- and microstructures along with their appearance and extinctions, its lawful nature as well as its randomness, which together sustain the drama of cosmic history and its processes.

Conclusion

In this chapter, we have seen that the outside story and the inside story are intimately connected, the former providing the necessary conditions for the latter. While the former sustains and suspends the latter between the two poles of existence and nonexistence, the latter manifests itself tangibly in outward expressions such as art, music, and religious expressions. The challenge is how to do justice to both stories at the same time. After all, the physical and chemical processes necessary for the moment of human consciousness have been at work from the beginning, providing the substrate for the emergence of subjectivity as a very real phenomenon in the universe. Its predecessors were the first flaring forth of the universe in the big bang, the galaxies, the stars, the solar system, the formation of our planet, as well as the evolution of life against all odds, culminating in the emergence of the human brain with its intricacies and mystery. The role of the brain in human experience is without equal; without it there would be no experience, no inside story, and no possibility for revelation.

---

49. Haught, *New Cosmic Story*, 68.

7

The Inside Story Solidifies—Religious Symbols

ONE OF THE MOST astounding human capacities is the ability to symbolize. Indeed, so significant is this capacity that we find it at the root of all our practical and intellectual achievements. It is based on our ability to perceive the world holistically, and without it we would be less than our closest animal cousins, the chimpanzees, for they possess a rudimentary capacity of this kind. The task of this chapter is to take our understanding of the evolution of human consciousness a step further by more fully exploring our ability to symbolize, then relating it to what is known as *transempirical reality* or religious experience. Such an attempt must steer a middle course between two extremes. On the one hand, we should refuse to elevate any form of religious beliefs to an ideological level that excludes all scientific data; on the other hand, we must refrain from an exclusive trust in science that rejects everything else as human projection or psychological instability.

On the Way to Revelation: Symbol Recognition

Symbols allow us to give outward expression to representations of inner experiences, and so symbol formation serves a twofold function: cognition and communication. Symbols enable us to live not only in the world of tangible phenomena but also in the invisible inner world of experience. Moreover, they allow us to inhabit the world of mental abstractions, parables, and paradoxes. Through symbols we can even ponder what it might be like not to exist. Coupled with the ability to formulate language, humans can build an almost infinite number of concepts and propositions, including fictional ones.

Symbolic behavior is not unknown in the animal world; it is especially common during the mating season. Many such gestures involve the giving of gifts and dancing. The male nursery web spider, for example, offers a small parcel of food wrapped in white silk to a female as an invitation to mate. Among birds, the monogamous great crested grebe bonds for life, and the pair will frequently conduct pair-bonding dances not only during their courtship but throughout their partnership, especially when raising their young. Dance routines vary with the circumstances in life. Prairie voles, tiny and very affectionate mammals, go even further. When one partner is stressed, the other will come and give the equivalent of hugs and kisses. Surprisingly, symbolic behavior does not require highly evolved brains. For instance, courtship rituals among insects, not known for extensive neural networks, are quite common. Dance and flight rituals may occur during courtship; males may stroke females with their legs or antennae; wings may be moved in circles, often woven into detectable routines, so that one may even speak of ritualization.[1] What stands out is that certain behavioral patterns have been mapped onto a new function without losing the original context. Feeding has been extended to aid pair bonding. Motor skills have been raised to the level of rhythmic dances to serve procreation. These animals have obviously assigned significance to certain behavioral patterns that carry a surplus value in the form of a message beyond the original function, while the rituals themselves stimulate the cognitive apparatus of the receiver to ensure perception of the message and so serve the process of natural selection.

Similar principles underlie human behavior, where postures, feeding, and caressing gestures are transformed into rituals. It is therefore likely that rituals such as these are more ancient than other ways of expressing human experience symbolically—for instance, in speech or narrative accounts. At the earliest human level, such rituals would have involved dancing, drumming, rhythmic vocalizations, as well as cave paintings and even ritual killings. All these nonverbal expressions carried

---

1. Smithsonian, "Mating in Insects," http://www.https://www.si.edu/spotlight/buginfo/mating.

symbolic messages that related to people's empirical world, including their inner world of experience.

Early modern humans, sometimes referred to as Cro-Magnons, a name derived from a rock shelter in southwestern France, appeared in Europe around 40,000 years ago. These hunter-gatherers populated eastern, central, and western Europe for 25,000 years, a period that included the last ice age (it peaked in Europe around 20,000 years ago). Known for having a larger brain capacity than all modern humans, they hunted big game (mammoth, cave bears, wild horses, and reindeer), built semi-permanent camps, and worked fine antler and flint points.[2] Except for one mesmerizingly new and puzzling element, all this would fit with the idea of a straightforward cultural advance from what had gone before. This period shows the sudden and mysterious appearance of an unprecedented mastery of sophisticated symbolic art, with which early modern humans decorated the walls of hundreds of caves deep underground across Europe and in Indonesia and Australia. Repainted by many generations over thousands of years during the Upper Paleolithic, these caves are often regarded by scholars as cultic sites.[3] Wild bulls, buffalo, bison, and bear occur most frequently and in prominent poses, expressing awe-inspiring vitality and power—symbols of the universal power of nature. For the Cro-Magnons, these animals were the incarnation of that power, through which they experienced a transcendent dimension of reality.

Prior to this period, humans had no art, no sophisticated symbols, no religious expression; now these appear in brilliant multicolor. What had changed? Although we will never know for certain, the spontaneous appearance of this uncanny ability, its flourishing over tens of thousands of years, as well as its vanishing 10,000 years ago, suggests that a major breakthrough in human consciousness had occurred: the recognition of a transcendent dimension of reality with far-reaching implications.

Perception of Symbols in Religious Experiences

In what follows, I explore the more specific case of symbol perception in religious experiences. To begin with, we need to keep two general features of symbolic perception in mind. First, the line of demarcation between the perceiver and what is being perceived is not as clear-cut as

2. Tuttle, "Human Evolution."
3. Lewis-Williams and Clottes, "Mind in the Cave."

it is in scientific experiments, simply because in religious experiences the perceiver is identical with the one who personally encounters what is being perceived, and all succeeding perceptions will be influenced by the earlier one. In other words, in religious experiences we are personally involved. Second, perceptions function on two levels. At the surface level, we encounter no more than an object or event, say thunder during an electrical storm or a cool breeze after a hot day. Yet often while perceiving such phenomena, another, holistic perception usually occurs in response to the first, which goes beyond what is merely functional or factual, typically involving acts of our imagination. And it is through this second, holistic perception that we can turn practically everything in the universe into symbols, including religious symbols, but this kind of perception only occurs when we open ourselves to the world existentially. While we remain uninvolved critical observers, this second-level holistic perception will elude us.

The English philosopher and theologian Ian Ramsey (1915–1972) grappled with the challenge of religious language in a contemporary context. Aware of the claim by scientists that religious talk is meaningless because its truth claims can never be established, as well as of the criticism that religion belongs to the realm of the unsayable and that therefore silence is the only valid response to religious experiences, Ramsey took a novel approach. Adopting the principle of personal disclosure as the basis for explaining religious language, he was able to take it beyond the merely subjective, because in language generally personal experience and disclosure are inseparable. Besides, all religious language is analogical; the religious words humans create will always involve the language of analogy and so will not be scientifically verifiable even when humans respond to what they claim is a revelatory experience. In short, Ramsey's disclosure model suggests that, as with the use of symbols, there is relevance in religious language, and that this language contains something more than a merely scientific description of human experience. Seen from this perspective, through second-level or holistic-symbolic perception, it is possible that a new dimension may break open that enables us to see, amid commonplace activity, our world in a new light.[4]

---

4. Ramsey, *Religious Language*.

The Nature of Religious Experiences

On the tiny Aegean island of Lispis, storytelling, common symbols, vernacular religion, hierophanies, saints, and miracles intermingle enduringly in daily community life, "even when the framework of official religion has changed," notes Marilena Papachristophorou, professor of folklore studies at the University of Ioannina (Greece). As an ethnographer, she has studied the island's narrative culture and oral tradition for more than a decade. Here the natural and supernatural worlds "are perceived as an indivisible whole whose parts are in constant communication, either through miracles, hierophanies, and visions or through an abundance of wishes and invocations that people utter all the time in their everyday routine."[5] Perceptions such as these unfold today as a matter of course and shape not only the understanding of the community's local history but also their collective identity and worldview. Such religious intensity may come as a surprise in an age of growing rationalism, but it becomes more plausible when we understand that the island of Lispis has always been linked to its neighbor to the west: the island of Patmos, with its Holy Monastery of St. John the Divine.[6] As we continue to explore symbolic cognition in human experience and religion, this contemporary example of the formative influence of religion on community and culture serves as a type for what in all likelihood took place in earlier phases of human history as our ancestors experienced their world more holistically than we now do.

Religious historian Mircea Eliade (1907–1986) distinguishes between religious and nonreligious experiences. Eliade collected religious documents that express religious experiences and then interpreted them, taking religious phenomena "on their own plane of reference."[7] He calls religious experiences "hierophanies" or manifestations of the sacred.[8] Eliade writes:

5. Papachristophorou, "Ordinary Stories," 61.

6. John the Divine, also known as John the Theologian, is the titular name conferred by tradition on the author of the final book in the Bible, the book of Revelation. Its text states that its author is known as John and, at the time of writing, lived on the Greek island of Patmos in exile, banished during the anti-Christian persecution under the Roman emperor Domitian (ruled 81–96 CE). Lispis's patron saint is the "Holy Virgin of Charou" (*Panagia tou Harou*).

7. Eliade, *Patterns*, xiii.

8. *Hierophany* is a word of Greek origin combining the adjective *hieros* (= sacred or holy) with the verb *phainein* (= to reveal).

> This word [*hierophany*] is convenient because it requires no additional specifications; it means nothing more than implied by its etymological content—namely, that something sacred is shown to us, manifests itself. One may say that the history of religions from the most elementary to the most developed is constituted by a number of important hierophanies, manifestations of sacred realities.[9]

To interpret hierophanies, Eliade employs two critical concepts: (1) the dialectic of the sacred and the profane and (2) the centrality of symbolism or of symbolic structures. The first allows him to distinguish religious phenomena from nonreligious ones; the second provides access to the meaning of sacred manifestations.

For Eliade, the sacred is irreducible and must be allowed to speak its own language. He refuses to comprehend its essence by means of any other discipline, be it physiology, psychology, sociology, neuroscience, art, and so on. To do so, he argues, would miss the one unique and irreducible element in the sacred. Therefore, the interpreter must suspend judgment on all nonevident matters and sympathetically participate in the experience of the person undergoing such an experience. In practical terms, what Eliade attempts to do is to reconstruct the situation of sacred manifestations and thereby capture the intentionality of the phenomena from the evidence available to him.

Before going further, let me clarify what Eliade means by "religion" and "the sacred." Religion is not necessarily a belief in God, or gods, or ghosts. Rather, religion means being occupied with the sacred or with the transcendent, which is distinguished from "the profane." He uses the latter term in a nonderogatory sense. The profane is simply nonsacred, not antisacred. Religious experiences, therefore, "bring a man out of his worldly Universe or historical situation, and project him into a Universe different in quality, an entirely different world, transcendent and holy."[10] More simply put, religious experiences involve a person in a radical break with all secular and profane modalities. They point beyond the relative, historical, and natural world of ordinary experience. This quality opens human consciousness to transcendent values, which, according to Eliade, is religion's principal function.[11]

---

9. Eliade, *Myths, Dreams, and Mysteries*, 124.
10. Eliade, *Mephistopheles*, 76, 78–124.
11. Eliade, "Structure and Changes," 366.

What then is the sacred? To describe it, religious scholars have used a variety of terms, some more common than others: the "wholly other," "ultimate reality," "Being," "absolute reality," "eternity," "divine," "transcendence," "the transcendent horizon," and so forth. Lutheran theologian Rudolf Otto (1869–1937) coined the famous phrase "*mysterium tremendum et fascinans*" ("an awe-inspiring and fascinating mystery," as referred to earlier),[12] while Douglas Allen, a scholar who has published extensively on Eliade, has taken a more contemporary turn and added such phrases as "meta-cultural and transhistorical," "transhuman," "transmundane," and "source of life and fecundity."[13] Again, experiencing the sacred is to undergo a religious experience or a *hierophany*. For Eliade, religious and nonreligious outlooks are not merely chosen dispositions but modes of being, as a passage from *The Sacred and the Profane* makes clear:

> The non-religious man refuses transcendence, accepts the relativity of "reality," and may even come to doubt the meaning of existence.... Modern non-religious man assumes a new existential situation; he regards himself solely the subject and agent of history, and he refuses all appeal to transcendence. In other words, he accepts no model of humanity outside the human condition as it can be seen in various historical situations. Man *makes himself*, and only makes himself in proportion as he desacralizes himself and the world. The sacred is the prime obstacle to his freedom. He will become himself only when he is totally demysticized. He will not be truly free until he has killed the last god.[14]

Yet, in his work Eliade remains purely descriptive. Rather than claiming that religious phenomena possess transcendent truth, he simply pays attention to what they do reveal: on the one hand a dichotomy between the sacred and the profane, and on the other that religious individuals seek to experience the sacred by getting beyond the profane. In other words, these individuals believe that something comes from somewhere else and shows itself to them, and that whatever appears to them is the sacred, while the profane is that through which it appears. From this Eliade concludes that historical reality is only significant insofar as it reveals the sacred. In this view, especially in relation to the new cosmology that

---

12. Otto, *Idea of the Holy*, chapter 4.
13. Allen, "Mircea Eliade's Phenomenological Analysis," esp. 176–77.
14. Eliade, *Sacred*, 24.

occupied us in earlier chapters, our universe would have no value except to the degree that it reveals the sacred.

If, according to Eliade, literally anything can serve as a medium for religious experiences, we may conclude that our ancient ancestors would have seen hierophanies everywhere. After all, at the dawn of human consciousness nature would never have appeared as merely natural. What we don't know, of course, is why a tree, a rock formation, or a landscape should have turned into a hierophany, or, for that matter, at any time ceased to be one.[15] Yet there is evidence in religious history that suddenly appearing phenomenon like a devastating flood or a lightning strike, or living creatures of great power like the wild bull or the lion, or phenomena of fecundity and fertility like the recurring seasons and childbirth have served as symbols for the sacred. Whereas we must be tentative about the inner structure of such experiences and the exact nature of what those who experienced them recognized as disclosure, we must assume that for them the experiences were real. If we followed Eliade and took account of hierophantic manifestations on their own terms, we would refrain from dismissing religious phenomena as merely natural.

On the Structure of Religious Symbols

The previous section made us aware of the paradoxical coexistence of the sacred and the profane. To grasp the nature of religious symbolism better, let us pay closer attention to this paradox. According to Eliade, the paradoxical element is the idea that finite natural objects like trees, rocks, landscapes, astronomical or other natural events, as well as living creatures should reveal something of the infinite and the transcendent, while they themselves remain natural and historical. The profane never *becomes* the sacred. What's more, because of such an experience, the one undergoing it becomes conscious of a surplus that now attaches to the object that mediated the experience. In the process of paying attention, the perceiver attributes religious, even ultimate value to it. This is the paradox present in every hierophany.

With wonderment and awe, Eliade refers to it as the "great mystery." He writes: "The great mystery consists *in the very fact that the sacred is made manifest*; for . . . in making itself manifest, the sacred *limits* and

---

15. Eliade, *Patterns*, 38.

'historizes' itself."[16] Then, with an undeniable Christian allusion, he speaks of the "paradox of incarnation." Because the sacred "expresses itself through something other than itself," through a natural phenomenon, the sacred never fully reveals itself. Hierophanies merely point to it.

Throughout history, all over the world, people have used symbols in a great variety of forms to give meaning to the perceptions they draw from religious experiences. Our earliest ancestors may have resorted to gestures, dances, and rhythmic shouting to give visual, auditory, and dynamic expression to what they experienced in their encounters with the incomprehensible mystery they met in their natural world. Once human consciousness could more fully grasp the paradox that the infinite should manifest itself in the finite, the absolute in the relative, the immaterial in the material, more abstract artistic symbols of great depth and sophistication would have emerged. Today, scholars recognize that symbols and pictures are among the most important devices for grasping religious meanings and practices. They also recognize that religious symbolisms mediate and make intelligible the idea and presence of the holy, strengthening the relationship of human beings with the realm of the sacred or with the transcendent—that is, with the spiritual dimension of existence.

In this light, religion is a system of signs and of sign-based meanings, grounded in an experience of the sacred. As a system of symbols, religion is an objective phenomenon that exists because human beings perceive that they do not live in a world "as they find it," but in one they must interpret and thus change. German theologian Gerd Theissen elaborates on this idea when he writes:

> By work and interpretation, human beings make their world a habitable home. Here the transformation of the world through interpretation does not take place through causal interventions in nature as in work and technology, but through "signs," i.e. through material elements which as signs produce semiotic relations to something signified. Such signs and sign systems do not alter the reality signified, but our cognitive, emotional and pragmatic relationship to it: they guide our attention, bring our impressions together coherently, and link them together with our actions. Only in a world interpreted in this way can we live and breathe.[17]

---

16. Eliade, *Myths, Dreams, and Mysteries*, 125.
17. Theissen, *Theory of Christian Religion*, 2.

These are useful extensions of Eliade's ideas. Their usefulness lies in their integration of religion and culture, providing us with a living context for exploring some ancient symbols of divinity before turning our attention to the central symbol of Christianity.

Ancient Symbols of Divinity

Sensitivity to symbols is present wherever we see ritualized behavior, even in the animal world. When certain behavioral patterns are ritualized, they are disconnected from the original context to serve another—for instance, feeding when used in courtship or mating rituals. A surplus value is recognized that goes beyond original functionality, and the same may be said for religious experiences.

Based on what we have seen in our exploration of symbolic perception and its extrapolation into the realm of religion, we can safely affirm that encounters between religions and cultures can be successfully negotiated only on the level of their sign systems or symbols. In other words, only symbols have the power to articulate the way a culture or religion understands itself from the inside—that is, from the perspective of prevailing beliefs and theological insights. While a dialogue between a polytheistic culture and a monotheistic one would be more demanding than between two polytheistic cultures, comparative studies in religious history have shown that sign systems belonging to locally confined deities often represent the same aspects of the divine, as Georg Baudler notes: "For example, the Ugaritic Ba'al, the Sumerian-Babylonian Tammuz, and the Egyptian Osiris are nothing other than local variations and accentuations of the same basic human experience of fertility, the growth of vegetation, and the overcoming of destructive natural forces." He continues:

> The innumerable mother-godheads and love-goddesses of historic religion—Annat, Ishtar, Inanna, Shams, Isis, Hera, Gaia, Demeter, Artemis, and Lakshmi—are different models of the same existential experience of maternal and feminine love, while the divine bull and dragons and different godheads of thunder and lightning reflect the experience of a thrusting ferocity, of wildness and of the blind elemental violence of fate.[18]

Such large-scale representational images of the elemental forces that enhance as well as threaten human existence convey the same experience

---

18. Baudler, *God and Violence*, 27.

as the monumental cave paintings of the Upper Paleolithic mentioned above. In this early period, it is unlikely that our ancestors would have abstracted these forces from nature. Instead, they would have regarded their all-surpassing power as objectively real within their own horizon of perception, albeit as a superior dimension that transcended the merely biological. When we look at these paintings as well as the clay figurines of the same period, we encounter the attempt to express this transcendent dimension artistically by means of symbolizing models.

In this light, it is perhaps not surprising to find that later developments should elevate one of the oldest symbols, the bull, to express the transcendent dimension. While the bull was also understood as a symbol of fertility, its destructive force was at least initially foregrounded. A powerful exemplar comes from the entrance to the Minoan palace of Knossos in Crete (from about 5,500 to 3,500 years ago). A wall relief depicts a raging bull charging an olive tree with savage force, threatening the tree's life. What we see is the sacred fascination of our ancestors with the forces of nature, which they experienced as uncontrollable, untamable, and destructive, causing suffering and death. By extension, this overpowering and incomprehensible fate was understood as the wrath of some divinity.

In a lengthy study of Near Eastern myths and ancient Semitic fairy tales, German diplomat-scholar Werner Daum has attempted to reconstruct the underlying "primeval Semitic myth."[19] According to Daum, the basic pattern of this myth lives on in religious feasts and rites in the Mediterranean region, in Greece, Crete, Africa, and Egypt, even in Hebrew and Islamic monotheism. Despite its many layers, the basic structure of this myth is best preserved in the religious tradition attributed to ancient (legendary) Sheba. Here there are three divinities. Two are benevolent, disclosing the qualities of calm waters, fertility, abundance, warmth, and growth (also associated with sunlight); the name of the third is "Enraged," a divinity of destruction who wields a club and whose symbols are a charging goat and a bull. The latter relates to events experienced by early Mesopotamian farmers as the disclosure of the sacred, in raging floods that destroyed human efforts to cultivate croplands. And although the benevolent divinities seek to oust him, they never quite succeed.

In Yemeni tradition, for instance, this recurring theme goes back to a more ancient view of divinity: a powerful wilderness god whose fury people tried to appease with the sacrifice of young girls, whom they

---

19. Daum, *Ursemitische Religion*.

sent into the wilderness or buried alive at the head of the Wadi to stop the floodwaters from reaching the cultivated land. In an ancient Arab parallel, this divinity appears as an ogre whose name is Il Muquah. This divinity is *Il* or *El*, the god of the mighty and terrible rainstorm.[20] In fairy tales, he appears as a kind of water demon who has "neither sons nor daughters. He had always existed and was not begotten. Everything belongs to him; he owns the land. His dwelling place is the wilderness far from the homes of man. He uses his great power at his pleasure for good and for evil. Although he is sometimes kind and helpful, his wild and destructive character prevails."[21] Later, *El* is the name of the father god in the Ugaritic tradition, whose sons are Jamm and Mot, one the god of furious waters and the other of deadly droughts.[22] Here too *El*'s symbolic animal is the bull. However, in a later Syrian depiction bull power appears to be broken, for the fertility god Ba'al is shown standing with one foot on the back of the bull and one on its head.

In short, from earliest times humans felt that divinities of this kind threatened and controlled their fate. With rising consciousness of future possibilities, however, a threatening transcendental horizon can only give rise to a deep sense of anxiety and fear that even auspicious circumstances cannot assuage. So strong is this perception that folklore often interprets a string of good news and good fortune as warnings of disaster. Here we find ourselves face to face with one key question in the history of human experience: "How have humans dealt with what appears as the overarching reality that relentlessly and indiscriminately threatens human existence?" To fend off these threatening powers, to render life more secure and forge meaningful pathways through the severity of these threats, three quasi-religious approaches stand out: sacrifice, heroism, and wisdom or altered states of consciousness. I elaborate briefly on these alternatives in the next section.

The Quest to Render Life Secure

Possibly the oldest way of trying to avert tragedy and misfortune was to appease the destructive powers by sacrifice. In the ancient conception of life, nothing was free—even good fortune had to be paid for sooner or

---

20. Daum, *Ursemitische Religion*, 80.
21. Daum, *Ursemitische Religion*, 46.
22. Daum, *Ursemitische Religion*, 188.

later. Offering a costly gift to appease the wrath of an overpowering and destructive deity made child sacrifice not just a widespread but an almost universal practice. In situations of life-threatening danger or at crucial junctions in the life of the community, people's most precious possession would be handed over to this ruling deity. This could mean sacrificing their firstborn sons or, if a kingdom was being threatened, the king's marriageable daughter. The expectation was that such sacrifices would secure the harvest, avert a flood, or spare the kingdom, with disaster prevention as the primary goal. Good fortune as well as predator gods demanded their share.

Such ancient demands are reflected in biblical texts that command the sacrifice of the firstborn, although these demands are often softened by provisions for redemption.[23] Also, under the principle of giving the best to God, the first portion of the harvest and the firstborn of the flock (preferably male animals) were to be sacrificed.[24] In the New Testament, we still read about "ransom payment" (Mark 10:45). In St. Paul's letters, we read about the "high price" with which the new community had been bought (1 Cor 6:20). Today, sacrificial thinking is still subconsciously at work, albeit in a sublimated form illustrated by the common phrase "there is a price for everything," rendering the idea of a free gift culturally repugnant.

Such a sacrificial conception of reality may well have its origin in primitive layers of prehuman experience: when a group of early ancestors fled a predator and escaped, except for one who was taken. Violent in

---

23. What is shown in the Bible as the test of Abraham's faith ("Take now your son, your only son Isaac, whom you love, and go . . . and offer him . . . as a burnt offering on one of the mountains of which I shall tell you" (Gen 22:2)), Baudler interprets as a vestige of Israel's Canaanite surroundings, a demand of the ancient storm and weather deity, *El*, and although in this case no a priori provision for redemption was given, Abraham offered a ram in the end (provided by God in the Genesis account) as a substitute. This is Baudler's view of the origin of sacrifice. French-American anthropologist René Girard has developed a different approach based on the theory that in archaic communities sacrifice served as a violence-curbing strategy grounded in what he calls the "primitive sacred," which was believed to have appeared in their midst when an arbitrary victim was slain in a spontaneous outbreak of collective violence (see René Girard, *Violence and the Sacred*).

24. The Old Testament has no general word for sacrifice but sometimes uses *qorban*, meaning that which was brought near. Other words are *zābah* (that which is slain) and *ôlâ* (that which goes up or is burned). Included under *qorban* were the nonblood offerings like the cereal offering and the first fruits. The principle of substitution is present not only in replacing the human firstborn with an animal victim but in the provision for the poor, who could offer up doves for a sin offering (Lev 7:5).

principle, this mindset rules out the possibility that human existence can be structured differently. Even in contemporary culture, the sacrificial mindset is alive and well. We find it in the esoteric literature where blood rites, even human sacrifices, are celebrated as founding acts of the world and of humanity. Jungian psychologist Wolfgang Giegerich writes:

> In the blow of the sacrificial axe manifests the strength of the human soul as it gives itself to its own origin.... In killing lies in the strictest sense the origin of humanity qua soul and spirit, in killing humans won the fundamental possibility of agency and with it the possibility for the birth of culture.[25]

Some scholars have argued that sacrifice is a fundamental constellation of human existence and that the task of religion is to provide society with sacrificial rituals as tools to externalize the human fascination with fear and violence and so keep its toxicity as far as possible from real life.[26] Austrian theologian Raymund Schwager believed that humanity needs scapegoats and hence sacrifices as a violence-channeling mechanism. He regarded Jesus of Nazareth as the perfect scapegoat who took the "sins of the whole world" into his sacrificial death.[27] And, like cultural anthropologist René Girard (whose theory of cultural origins we will explore in chapter 8), Schwager saw Jesus' act of voluntary self-giving as the perfect sacrifice that exposed the archaic sacrificial religiosity as an illusion because it was itself structured by religious violence.

Another path humans have pursued to escape the arbitrary and life-threatening powers was the path of heroic acts. Although they are often intertwined with the sacrificial path, heroic acts represent a counterpoint to it when they serve as an all-out effort to overcome the forces that mar the pursuit of happiness. When our early and powerless ancestors, equipped with little more than wooden sticks, faced down giant mammoths, snarling cave bears, and wild bison, what seems to have motivated them was more than gaining access to food. It was the metaphysical desire to symbolically conquer (in a magical way) what threatened them, thus turning the kill into a cultic-heroic act. Vestiges of such archaic heroism are still visible today in the Spanish bullfight because the matador does not primarily fight the bull; he symbolically fights the power of death. If our early ancestors ate a portion of such a kill and sacrificed

---

25. Giegerich, *Tötungen, Gewalt aus der Seele*, 47, my translation.
26. Gutmann, *Die tödlichen Spiele*, 192–94.
27. Schwager, *Must There Be Scapegoats?*

the remainder, they accomplished a dual purpose. Through the power to kill a more powerful animal, they raised themselves at least to the same level as the forces they were seeking to overcome, if not to a higher one. And to the extent that the slain animal was also a sacrificial offering to these powers, by consuming part of it they were symbolically eating at the same table with them. In this way, early humans had reached beyond their puny status; they had become predator gods themselves.

A variant of this heroic posture appears in Greek mythology in the figure of Prometheus. He knows that the gods are mightier than he is, but this knowledge does not lead to his submission. Instead, in titanic spite, he rebels. The gods can chain him to mountains and torture him eternally, but they cannot break his spirit. With his heroic act Prometheus grasps a freedom that even the gods cannot vanquish.

The third path worth mentioning is the path of wisdom based on altered states of consciousness. One of the primal responses to existential threats is fear. With rising consciousness of our finiteness and the inevitability of death, fear is heightened, leading to the endless creaturely struggle for the necessities of life. The path of wisdom turned this struggle around by declaring that any attempt to render life secure is illusory. In Buddhist thought, for example, the thirst for life, which is the root of all anxiety and creaturely suffering, expires to the extent that the meaninglessness of this striving permeates our consciousness. Those who have realized this wisdom in the depth of their being are enlightened, wherein they find freedom from the fear of adversity, sickness, and death.

At the same time, Buddhists know that it is not wanton asceticism—where humans tear themselves away with all their might from their desires—that leads to enlightenment. Buddhist wisdom suggests that deliverance from the thirst for life is not achieved by self-assertion in strenuous ascetic efforts (not unlike Promethean heroism) but by surrendering to what is given. Not high-flying abstraction, but quiet contemplation and beholding bring mystical insight into the necessity of aging and dying, just as the patient internalization of this insight renders the thirst for life increasingly inoperative. In other words, accepting everything life presents, allowing oneself to be gifted, even enchanted by it, lies at the heart of the path of wisdom. In the most general sense, this path is found in the pages of Buddhist writing, in the sayings of Confucius, in the work of Lao Tzu, and in Platonic dialogues.

While similar thoughts have appeared in certain branches of contemporary philosophical practice, they foundered on Nietzsche's elementary questions: "What is it that I am doing?" and "In doing it, what am I after?"[28] Answering them led in the final analysis to the dilemma already recognized by Gautama Buddha: everything I do aims at the enhancement of life, which in light of the inevitability of aging and death is eventually meaningless. This paradox of the meaninglessness of existence necessarily persists, as Nietzsche clearly saw, unless death is not the last word. With this insight we have reached the point that leads logically to an exploration of the biblical path that (like Judaism and Islam) overcomes death.[29]

So far in this chapter, we have traversed a vast stretch of religious history. Beginning with the notion of symbol recognition, we recounted how with rising consciousness our ancestors conceptualized the arbitrary and life-threatening forces of nature that accompanied their existence and how they sought to overcome them to render their lives more secure. This journey took us from bull power to the (nonreligious) path of wisdom—which, as we saw, ends in an unresolvable dilemma—to the cusp of something new in religious history, the rise of monotheism.

The Rise of Monotheism

Monotheism did not fall ready-made from the sky; rather, it emerged in religious history over time. Monotheism made its first appearance in Egypt around 1350 BCE, when the "heretic Pharaoh" Akhenaten, known for his unconventional devotion to the sun, proclaimed Aten, the god of the sun disc, the sole deity and supreme ruler of the universe. However, his attempt to eliminate the other gods was unsuccessful, and after his death everything reverted to the old pantheon. Two hundred years later, monotheism resurfaced in Iran in the teaching of Zarathustra around 1100–1000 BCE. In a vision of blinding light, so the story goes, Zarathustra—also known as *Zoroaster* to the Greeks—stood in the presence of an unfamiliar god, who revealed himself as the sole god of the universe. He had no name, but Zarathustra called him Ahura Mazda, the "Wise Lord."[30] The significance of this event for religious history was not only that a monotheistic religion reappeared but that this unfamiliar deity seems to have had a new relationship with human beings, for Zarathustra,

---

28. Nietzsche, *Also sprach Zarathustra*, 399.
29. Baudler, *Die Befreiung*, 31, 40, 44.
30. Lommel, *Die Religion Zarathustras*.

in encounters with Ahura Mazda, received words of *revelation*. Based on these revelations, Zarathustra presented Ahura Mazda as the source of human morality and as the god who would judge everyone by their thoughts, words, and deeds after they died, although "morality played no part in how one experienced the afterlife."[31] Zoroastrianism became the official religion of ancient Persia in the seventh and sixth centuries BCE. From the perspective of human cognition, however, there was a huge problem: deities like Ahura Mazda were abstractions. How were worshippers supposed to connect with them when they had no human form and attributes?

Several centuries after Akhenaten and broadly coeval with Zarathustra, a small group of Semitic tribes, later known as Israel, faced a similar problem, as the first successful monotheistic religion arose in several stages out of their midst. Since its rise is closely tied to Israel's history, this movement is best related to four decisive periods of that history: from early pastoral/tribal Israel to the monarchy or the pre–First Temple period (about 1200–1000 BCE), from the monarchy to the exile or the First Temple period (1000–598/6 BCE); the exile or the Babylonian period (598/6–538 BCE), and finally the postexilic or Persian and Hellenic period, from 538 BCE to the destruction of Jerusalem in 70 CE.[32]

The religious picture in early pastoral Israel was quite diverse, although the prevailing polytheism contained monotheistic elements in the form of monolatry, whereby one deity was given regionally higher status. The deities worshipped were *El, YHWH*, Baʻal, Gad, Anath, Zedek, Asherah, *El Elyon*, and others. Each region had their local deity. While religious devotion by families or clans was often centered on a single deity, this form of religion must not be mistaken for monotheism, as these deities functioned as mediators to a higher god. It is possible that *YHWH* was first worshipped exclusively in Gibeon by a small minority, yet in cultic expression this early Yahwism would have been indistinguishable from any other West Semitic religions.[33]

31. Aslan, *God: A Human History*, 99, 248–49.

32. Based on Gnuse's "Emergence of Monotheism," I have rearranged and synthesized the views of seven Old Testament scholars—John Smith, Gösta Ahlström, Valentin Nikiprovetsky, Fritz Stolz, Hermann Vorländer, Norbert Lohfink, and Gerd Theissen.

33. The tetragrammaton or *YHWH*, the name of the God of Israel, occurs first in the second creation account (Gen 2:4). He makes his first appearance in Exodus in the burning bush, where he speaks of himself as the God Abraham, Isaac, and Jacob (Exod 3:6), but paradoxically declares in Exodus 6:6 that Abraham did not know him by that

There is archaeological evidence that *YHWH* was served by various other deities, including Asherah, his female consort. Yahwism rose to some prominence under Saul. His successor, David, having chosen *YHWH* as the deity of his dynasty, elevated *YHWH* to the position of Israel's national deity. Local deities were beginning to be merged with *YHWH*, like *El Elyon* of Jerusalem and Ba'al of Bethel.

During the monarchy the royal court took responsibility for religious affairs: each king determined the national deity. *YHWH* worship remained a minority position that conflicted with the common familial religious expressions, sometimes opposing the religion of the court. There is some scholarly consensus that during this time the minority group began to write the biblical texts, projecting them onto the distant past. While the classical prophets and the later Reform under Josiah (620 BCE) sought to bring Yahwism to the masses—Hosea (750 BCE) is seen as the first *YHWH*-only prophet—monotheism remained the minority view, even down to the time of Jeremiah (580 BCE). Some scholars see preparatory stages in the emergence of monotheism, like the split of the nation into Israel and Judah under Jeroboam (930 BCE). If *YHWH* could function as a national deity of two nations, then a more universalist conception was within reach, aided by the notion that *YHWH* was imageless. Moreover, the oracles of the eighth-century prophets Amos, Hosea, Isaiah, and Micah attacked polytheism in the cult and further developed universalist ideas, so that Jeremiah's prophecies could imply that there was only one deity. With Second Isaiah the exclusiveness of *YHWH* came clearly into view, providing the defining element of monotheism, the *moment of exclusion*. Temple sites from the monarchic period have been discovered at Arad in Judah (dedicated to *YHWH* and Asherah);[34]

---

name, only by the name *El Shaddai*. This must have been confusing for the Israelites. There is no consensus about the origin and etymology of *YHWH*. The oldest plausible view (based on an Egyptian inscription from the fourteenth century BCE) relates *YHWH* to the land of *Shashu*, whose inhabitants were nomads from Midian. This hypothesis that *YHWH* was originally the name of a Midianite desert and weather deity makes sense in light of other data—for instance, the role of Midian in the Old Testament (Moses had married into a Midianite family and his father-in-law, Jethro, was a Midianite priest), the thundering phenomena that accompanied the meeting between *YHWH* and the Israelites at Mount Horeb, and the absence of *YHWH* from Canaan, the land of Abraham's sojourn. The Canaanites worshipped *El*, whose name governs the first creation account (Gen 1:1–2:3), also identified as the name of the deity Abraham knew (Gen 6:3).

34. Negev and Gibson, "Arad (Tel)," 43.

others include a Yahwistic temple at Elephantine in Egypt[35] and one at Tel Motza near Jerusalem.[36] In sum, then, it took 600 years for the exclusiveness of *YHWH* to become established in Israel, although true monotheism was still in the future.

It took the crisis of the exile for all of Israel to become committed to monotheism as a total cultural and religious system. Surrounded by foreign religions, Israelites preserved their religious and ethnic identities by not participating in foreign cults. The exclusive worship of *YHWH* fostered religious and social values, so that monotheism became connected with an egalitarian society, with the struggle for justice, and with the emergence of exilic literature in which old oral traditions were drawn together. The Pentateuch (the first five books of the Bible) was created in that period (some scholars even place it in the postexilic period) together with the historical narratives, while monotheistic assumptions were projected back onto Israel's early history. Since the exiles were mainly upper-class Jews, the intelligentsia began to interpret *YHWH* as the creator of the world and the divine guide of human history; as a result, the old national deity of Judah became a personal deity for many Jews during that period.

During the postexilic period, priestly laws that stressed pure monotheism and a new social order separated Jews from others. In effect, only postexilic Jews were pure monotheists; they portrayed God as the distant creator of the world but also as a close personal deity. From the Persian-Hellenic period onward, perhaps in reaction to emerging philosophical monotheism in Greece and Persia, Jews and later Christians began to declare their monotheism as the unique revelation and manifestation of God.

The Biblical Trajectory

In the first chapter of Genesis, the biblical God shares creative power with creaturely realms: "Let the waters bring forth swarms of living creatures . . . " (Gen 1:20) and "Let the earth bring forth living creatures . . . " (Gen 1:24). As the biblical path unfolds, the theme of power sharing is further

---

35. http://www.ancientsudan.org/articles_jewish_elephantine.html.

36. http://www.antiquities.org.il/article_eng.aspx?sec_id=25&subj_id=240&id=1975. Political and religious sensitivities have prevented excavations at the Temple Mount in Jerusalem, so that there is no evidence for the existence of the first temple from extrabiblical sources.

elaborated toward liberation from any form of victimization; more than that, the biblical God is perceived as being on the side of the poor, the needy, and the marginalized, becoming even like them in Jesus Christ. This counterintuitive story, to which Israel's growing awareness of vulnerability had led it, undercuts all aspirations to ancient forms of self-salvation via other paths, whether Promethean defiance, cultic heroism, or altered states of consciousness. But if God is on the side of the weak, the vulnerable, and the victims of history, the spirit of the biblical path seems to run against the grain of biological evolution and its harsh selection pressure. What should we make of that? First, some biblical facts.

The biblical account of Israel's history begins with a rescue operation. An enslaved group of people is delivered from oppression through specific actions of *YHWH*, the God of Israel. In Israel's imaginative remembering, this God had revealed himself to Abraham (although under another name; see Gen 6:3) several hundred years earlier as a deity with whom humans could dialogue—a novelty in religious history. This God was quite unlike the other deities of the ancient world: approachable, speaking with Abraham as with a friend, seeking a response from the one he addresses. Abraham, in turn, responds with reverent openness and obedience. More astonishingly still, openness and responsiveness rather than sacrifices and ritual seem to be the only preconditions for the human-divine relationship. In other words, the biblical path opens out into a breathtakingly new transcendent horizon.

When we look at it closely, this dialogical principle is already imaginatively folded into the creation story itself, as the text presents us with a richly interactive account. Creation not only responds to divine utterance, but the Creator responds to what he has created. He evaluates and even learns from his observations—"It is not good that man should be alone" (Gen 2:18)—and confronts the otherness of what he has created through further shaping and blessing. The divine "Let there be . . . ," which occurs 11 times in Genesis 1, grants creation a certain autonomy and participation in the creative process, so that creaturely agents become involved in the unfolding and ruling of creation. By thus self-limiting God's own sovereignty, God opens the space for the freedom creaturely realms need to foster life-promoting relational interdependencies.

We encounter this principle more fully with Moses, to whom this God mysteriously appears in the burning bush (Exod 3:3), even revealing (at Moses's request) God's name, *YHWH* ("The God Who Is There" or "I Am Here"). Moses is charged to liberate the people of Israel, who are languishing as slave laborers in Egypt. When Moses reluctantly accepts the task (after some bargaining with *YHWH*), God, in collaboration with Moses, outmaneuvers Pharaoh, Egypt's ruler and archetypal symbol of enslaving state power. Israel's liberation is not the result of a slave revolt against their masters, but the faithful fulfillment of God's covenant promise made to Abraham. The "I Am Here" God now leads the people of Israel through the Red Sea and on a long journey to the promised land, coming to their aid when they are in trouble.

Israel's memory of this journey speaks eloquently of the difficulties the acceptance of this counterintuitive image of *YHWH* posed as the transcendent horizon in Israel's collective consciousness. From an external viewpoint, Israel's original social structure seems to have differed little from the surrounding culture. Yet from *within* this often threatened and brutalized nation, taken captive by the superpowers of Assyria and Babylon, something quite extraordinary emerged: a body of sacred writings that testifies to a deep religious life.

Even as their monotheism evolved, the writings of prophets, poets, and sages criticized the reliance on military strength, rebuked the use of violence to maintain social cohesion, and challenged the people of Israel to instead trust the "I Am Here" God, whom they had perceived from the beginning of their journey as caring and protective. These texts also testify to struggles with life's exigencies that bore the stamp of personal and social destinies under the hand of *YHWH*, whose actions were reflected in what they deemed to be their national history.

However, this narrow conception too would undergo a vital transformation in the crucible of the exile as *YHWH* emerged as the God of all peoples, so that, as far as Israel was concerned, there were no other gods. Paradoxically, the loss of political identity and freedom had universalized Israel's transcendent horizon. In their perception, *YHWH* had chosen them from among all the peoples to testify that there is only one God; not only that, their latest catastrophic historical experience was interpreted as *YHWH*'s apocalyptic judgment, from which, in the end, they expected that *YHWH* would save them.[37]

By the time of the New Testament, Israel's transcendent horizon still resounds with the voice of judgment: John the Baptist addresses his audience as "brood of vipers" and declares that "even now the axe is laid to the root of the trees" (Matt 3:10). Yet a fresh note seems to accompany this

---

37. Theissen, *Biblical Faith*, 90.

voice "in the wilderness" that calls the people to avert calamity through repentance and baptism as a sign of their return to Israel's God. When Jesus appears on the scene, although he too preaches repentance, the accent he sets is even more surprising: God's rule is dawning, and the God Jesus proclaims is none other than the presence of his *Abba* (Aramaic for "father").

Jesus and the Rise of Abba Consciousness

Jesus himself lived and worked out of the vision that the first signs of God's "kingdom" (Greek *basilea*) were already appearing.[38] His message invited the "lost sheep of Israel" to turn around and participate in the embryonic appearance of *YHWH*'s commonwealth, made visible in prophetic signs—for example, in Jesus' table fellowship that embraced the marginalized and the outcasts. While official resistance to him showed how profoundly *YHWH*'s caring nature had been forgotten among Israel's religious leaders, the common people heard Jesus gladly, although they often misunderstood him when he radicalized elements of Israel's traditional norms. True, annihilating judgment may still be threatening (*mysterium tremendum*), but a second chance was opening as a sign of the enduring goodness and faithfulness of Jesus' *Abba*, the "I Am Here" God of the Old Testament. After all, was not the sun still shining on the evil and on the good? God is for life; hence life must be given the opportunity to flourish: "I came that they may have life, and have it abundantly" (John 10:10).

In Jesus' proclamation, judgment has already (partly) taken place: he sees the enslaving powers fall like lightning from heaven (cast down from the realm of power) and God's commonwealth coming to those who received this message with a good heart. As Gerd Theissen points out, Jesus, like John the Baptist before him, preaches judgment and salvation in the same breath, but with a fresh emphasis. The proclamation of judgment anticipates the coming of the "son of man"—the eschatological judge—who will not accuse but stand beside those who know that they must depend on God's grace, while the anticipated victory over evil

---

38. *Basilea* = "kingdom" but may be better rendered "commonwealth," as Jesus' *Abba* does not rule in a monarchical manner.

changes the proclamation of God's commonwealth into an offer of God's merciful rescue.[39]

This offer forms the background to Jesus' ethical teaching, in which he universalizes the entire Jewish law (Torah) by declaring that *loving God and one's neighbor as oneself* is all that is required. What Jesus preaches is not a new form of enslavement, but the perfect freedom entailed in the proclamation of a new social order where neighborliness is not limited to our closest circle but widened to include all those in need of God's mercy. Moreover, Jesus also radicalizes the demands of the Torah (the first five books of the Jewish scriptures): anxiety over existence only leads to inner enslavement, but when one trusts God one can give away one's goods; anger leads to murder; purity is an inner attitude rather than outward observance; erotic fascination with a woman other than one's wife is adultery; many words in prayer are superfluous because God knows the situation; meeting urgent human and animal needs fulfills the Sabbath law. Traditional Jewish social norms are equally reversed and relativized: for adults to see the *basilea* of God, they must become like children; tax collectors and prostitutes will enter the commonwealth before pious Jews; foreigners will eat with the patriarchs; and more.[40]

How shall we explain this quantum shift in Israel's transcendent horizon issuing from Jesus' proclamation? One likely answer is that Jesus does two things simultaneously: deliberately tossing the entire system of traditional norms beyond human reach, instantly universalizing human culpability before God. From now on, everyone who judges his neighbor must first take the plank out of his own eye before attempting to remove the splinter from the eye of another. In other words, true judgment belongs to transcendent reality.

In word and action, Jesus makes another universalizing move. He models his own intimate relationship with the Creator before us. As a Jew, he speaks with God and about God in a Jewish manner but—on a personal level—he speaks with God and about God as his *Abba*, a term so intimate that it repeatedly baffled those who heard him use it. The Greeks translated this Aramaic word to mean "father." This translation has, no doubt, thrust the idea of a father God into Western culture. While "father" is technically correct, in the original language *Abba* conveys a far more nuanced meaning. It can also mean "parent," "ancestor," or

---

39. Theissen, *Biblical Faith*, 91.
40. Theissen, *Biblical Faith*, 92–95.

"founder." As one Aramaic scholar puts it, "The root points to all movements in nature and in the cosmos. . . . This root also helps form the root of one of the words for love in Hebrew and Aramaic [with the implication that] parenting in the cosmic sense [is intended] . . . which is beyond gender . . . indicating a process . . . that begins in unity and gives birth to new forms."[41] When, therefore, Jesus speaks of and to his *Abba*, we cannot help but sense an intimate tenderness toward the one he addresses because Jesus knows in his own experience the all-embracing arms of the divine Parent. By the same token, Jesus is also deeply aware of the womblike quality of his *Abba*'s love. It was this *Abba* that Jesus came to embody and whose love he sought to awaken on Earth in a new and special way. According to Paul, this is the path of "faith, hope, and love" (1 Cor 13:13). It differs decidedly from the ancient paths noted earlier: sacrificial appeasement of transcendent powers, heroic self-assertion over and against the arbitrariness of fateful circumstances, and the attempt to meet threatening aspects of life with altered states of consciousness. By contrast, Jesus' path, if embraced fully, leads out of the age-old entanglement with violence and death to new ways of living through the one who remains hidden behind the pages of Scripture and behind all that exists.

However, one further point remains to be made. Active participation in the life dedicated to following Jesus' path is fraught with its own hazards. After all, the hardships and heartaches of human existence do not disappear. Instead, we are invited to face them with courage and humility, surrendering our cares to the "I Am Here" God, trusting the promise that, as we live through them, we will be met with the inner sustenance and comfort from the font of the universally available transcendent horizon, which, according to this path, is the all-sufficient presence of the divine Parent.

Christ: The Revelation of God as Love

The word *Christ* is a title that describes an appointed function. Derived from the Greek verb *chrio* (χριω), meaning "to smear or anoint," it designates a practice by which people were inaugurated or appointed to a specific office in ancient Israel by the ritualistic act of anointing. For instance, a high priest, a prophet, or a king would be anointed into office. While

---

41. Douglas-Klotz, *Hidden Gospel*, 18–19.

christos (χριστος) literally means "the anointed one," strictly speaking it should mean "the inaugurated one."

In Christian theology the word *Christ* refers exclusively to Jesus of Nazareth. This nexus, however, in no way exhausts its meaning. For, in its most general sense, this symbol means that this one crucified and risen man is also God experienced as graciously incarnate in the world. In other words, the crucified God, the Christ, is the central Christian symbol.

Delving deeper into the meaning of the symbol, we encounter an almost incomprehensible reality. The God who meets us in the crucified Christ seeks to draw us into communion with him: we are to live *in God* and *from God* or, as the New Testament has it, "in him live and move and have our being" (Acts 17:28). Since the crucified Jesus belongs to the divine identity, two profound implications emerge for our understanding of God as well as for our understanding of human existence, of the world, and of history.

Regarding our understanding of God, the symbol communicates what is beyond the grasp of our natural or even our religious understanding: God as creator and sustainer of all things would have taken up into his own identity all the contrariness of the world—every humiliation, every injustice, every sin and betrayal, every aspect of suffering, even death. More than that: this unfathomable reality is rooted in God's love, which descended in Christ to the lowest level of human nature and death in order to take us to a level of being we cannot reach by human efforts. In other words, the symbol means that the God who revealed himself in Jesus' despised and rejected humanity, addressing us in the symbol of the cross to confront us with ourselves, yet as accepted by "the boundless suffering love of God,"[42] continues to exert an extraordinary pull on human hearts. Gil Bailie has called it "the gravitational power of crucified love."[43] Or in Moltmann's words:

> This [love] has far-reaching consequences: religious desire for praise and might and self-affirmation are blind to suffering—their own and that of others—because they are in love with success. Their love is *eros* for the beautiful, which is to make the one who loves beautiful himself. But in the cross and passion of Christ faith experiences a quite different love of God, which

---

42. Moltmann, *Crucified God*, 213.

43. In his *Violence Unveiled*, Gil Bailie refers to Jesus' words in John's Gospel, "When I am lifted up [not to heaven, but on the cross], I will draw all men to myself" (John 12:32).

loves what is quite different. It loves what is sinful, bad, foolish, weak, and hateful, in order to make it beautiful and good and wise and righteous. For sinners are beautiful because they are loved; they are not loved because they are beautiful.[44]

The great irony of our time is that a post-Christian milieu casts God in the role of the great oppressor, the "cruel huntsman" of Friedrich Nietzsche's conception, who torments the human conscience with guilt and fear of eternal damnation. But the symbol of the cross properly understood tells a different story. The God we encounter in the crucified Christ does not hunt us with cruel intent but seeks to lift us even into the divine life itself. In those who are willing to open themselves to an understanding of the symbol as the fullness of divine love, unavoidably a new consciousness dawns that there is an answer to ultimate human longing. And this answer lies neither in religious observance nor in morality but in the image of God mediated by the Spirit of the crucified and risen Christ.

For the wider context of the book, it is important at this point to enlarge the aperture of our lens to include Christ's cosmic dimension. Far from limiting the experience of the cross to its one-off historical occasion, biblical understanding does not hesitate to link Christianity's central symbol to the history of creation from beginning to end: it speaks of "the lamb slain from the foundation of the world" (Rev 13:8) and, as "the alpha and the omega, the first and the last," also implicitly sees in him the ground of a new creation (Rev 1:11, 17; 2:8; 22:13). From this angle, the symbol may also be regarded as a symbol of God's history. While not identical with cosmic history, the symbol nevertheless interprets the history of the world and infuses it with both this-worldly and transcendent hope.[45]

The Depth of the Inside Story

In chapter 6, I advanced the idea that ever since the stuff of the universe became self-reflectively conscious in humans, cosmic history needed to be understood in terms of two stories, an outside and an inside story. There is the outside story of historical epochs and events governed by the laws of nature, investigated and described by scientists as uninvolved observers. Then, since the arrival of the human mind, there has also been

---

44. Moltmann, *Crucified God*, 213–14.
45. Moltmann, *Crucified God*, 218–19.

an inside story, hidden in the first since the beginning of time, a story composed in human consciousness by the subjective world of experiences and imaginings. I also noted that with rising consciousness a natural inclination emerged toward a search for a meaningful path through ever-present threats from unpredictable powers beyond human control. For our ancient forebears this quest was naturally religious. In their conception, everything they encountered was imbued with a mysterious presence, a sacred life force, that could only be communicated in symbols. Although biologically conditioned, this realization had dawned on them as part of the awakening to consciousness that had been on the way since the birth of the universe. Just as symbol recognition had emerged with the rise of the human mind, so religious consciousness appeared by virtue of the same mind.

Certainly, this awakening had been slow in coming in the evolving universe, and its gradual appearance, like all other events in cosmic history, had occurred in the wake of countless evolutionary steps and innumerable near misses, for the arrival of consciousness is as indigenous a part of cosmic history as the formation of galaxies, of stars and planets, of water molecules, and of bacterial life in hot deep-sea vents.

The events I have just mentioned were totally unknown when the first Christian communities formed around the claim that God had revealed himself in Jesus of Nazareth. The apostles and New Testament writers knew nothing of big bang cosmology, star formation, and evolutionary biology. Their religious imagination was informed by the "flat Earth" cosmology of the time. There was Earth beneath their feet; sun and moon circled the Earth, which was the center of the universe; above the lunar and solar orbits were the fixed stars located in concentric spheres around the center; and still higher up there was heaven, the dwelling place of God and his angels. This world picture informed not only their religious understanding, but also their conceptions of God, of creation, and of the meaning of human existence. This cosmography served as a solid scaffolding for their faith, worship, and theological imagination. Over time, however, it entered so deeply into the Christian conception of the world that it was regarded as an integral part of Christian orthodoxy. Subsequently, attempts to replace this outdated cosmology with a more current one were treated as an assault on the foundations of the faith.

Even today, the findings and implications of modern science, of deep time, and of evolutionary cosmology and biology have yet to enter mainstream Christian awareness. Most church leaders and theologians, let alone the broader community of believers, have yet to engage with the question of how to think of God in ways commensurate with the vastness of cosmic and biological processes as discovered by science. One of the premises of this book is the conviction that a good look at the evidence should convince them of the urgent need for a contemporary understanding of creation if Christians want to regain a measure of cultural competence.

Yet there is hope. Some scientists and scientist-theologians have ventured courageously into these uncharted waters. Believing that the depth of God's revelation in Christ is deeper than scientific models, these intrepid pioneers have begun to think of God outside tradition-imposed frames and of creation in terms of an unfinished, evolutionary universe. For us, there remains at the end of this chapter the task of braiding multiple strands of factual and hypothetical thoughts into a meaningful conclusion.

Conclusion

Earlier, I identified the epic journey of the cosmos as a movement propelled forward by yet-unrealized potentialities of being. This movement, I proposed, is at one level governed by the intricate interplay of physics and chemistry in all their multitudinous manifestations as observed by science, suggesting at the same time that this outside story is incomplete. I also suggested that as the cosmic story progressed a new, mysterious dimension emerged after billions of years. In the form of *Homo sapiens*, the universe had begun to think about itself and even sought to make sense of existence as such, as humans raised questions like "Why is there something where there could be nothing?"; "How can a meaningful course be charted amid ever-present existential threats?"; and "How can a sense of permanence be achieved in a perishable world?" In other words, the emergence of self-reflective consciousness launched the search for an integrative center of existence, a search that lies at the heart of the religious quest. Hand in hand with this development, humanity's growing cognition entailed the ability to symbolize, which empowered them to assign meaning to natural phenomena, vastly expanding their religious sensibilities.

With the rise of monotheism in Israel, the religious quest became more and more unified and personalized as it gathered religious attention around a single deity, whom they claimed had made himself known to them and called them into a dialogical relationship. This revelation was progressively refined until 2,000 years ago, or very recently on the cosmic calendar, when one unique human being—Jesus of Nazareth—appeared whom Christians believe embodied God and who in this *Abba* consciousness unveiled the source and goal of all creation: the unfinished cosmic journey, as death-defying crucified love. In this conception of reality, the primal character of the universe's inside story is not existential anxiety in a perishable world, but the implicit promise that this love will have the last word. This promissory character of the inside story, in itself a reflection of the desire of the one who promised, has been at work from the beginning, infusing the creation with longing for fulfillment, grounding all our striving for self-transcendence and more abundant life, even life beyond the grave. Yet, from the start, the path that leads to life is steep, walking against its gradient is hard, and the door is narrow.

The obstacles do not present themselves primarily as external circumstances but as rooted in our human constitution as highly imitative creatures. Thus, for all its effectiveness, our innate capacity to learn is underdetermined and therefore ambiguous, rendering the inner structure of human consciousness conflict-prone and tragic. Since this latter bias tends to inflect all we do, as we will see in the next chapter, it has even infiltrated the way we conceive of God.

8

Desire, Sacrifice, and Revelation

IN THE CLOSING SECTION of chapter 5, we explored the legacy of our evolutionary past, referring to it as our reptilian, mammalian, and neo-mammalian brains. I suggested that if left unaddressed some of our inherited proclivities will produce destructive tendencies that can threaten the well-being and even the survival of life on Earth. As it is, our world is reeling from disorder and violence. Political processes are unraveling, families are fragmenting, pollution drives the biosphere toward collapse, and any sense of historical responsibility seems to be disappearing. Our leaders find it hard to maintain even the appearance of political decorum, and the rest of us seem as baffled as they are by the mystery of the current crisis in its escalation toward extremes.

Interestingly, more than a century ago the French sociologist and father of academic sociology Émile Durkheim drew attention to a curious social phenomenon: that societies disintegrate when social distinctions are erased. Against this threatening observation, humanists assure us that we humans are equipped with consciousness, reason, and free will, and thus with the capacity for autonomous agency—capabilities that throughout history have enabled us to create adaptive solutions along with the most exquisite cultural artifacts. Yet, who will deny that humanity's present condition is markedly less orderly? We fly to the moon but keep fouling our nest in ways that defy all reason. Not only is our social fabric in serious trouble, but so is the life system that supports us.

Besides, we are a paradox to ourselves as we encounter within us two contradictory potentialities: potential for abyssal dysfunction (whose effects bombard us daily through the media) as well as the vast potential for authentic compassion and selfless love. A mystery indeed—and it is the task of this chapter to shed some light on its origin, history, and future.

Desire as the Basis of Human Agency

The one scholar who connects the dots for me more than any other is the French American cultural anthropologist René Girard. Girard is a thinker for our time. He stands fully in the postmodern milieu, yet also within the biblical tradition, as he brings these two strands into profound conversation. Going beyond earlier theorists of social disintegration, he identifies as its cause the very mechanism that he believes is also the basis of all human agency and social interaction. For more than forty years, in a dozen books and numerous articles, Girard has offered a captivating set of analyses that have much to say about the mystery we are to ourselves: a species with high aspirations toward transcendence yet unable to produce social stability and peace. His widely acclaimed work has produced groundbreaking insights into several highly topical issues: the passage of our species from "animality" to "humanity," the origin of archaic religion, the foundations of culture, the origin and role of sacrifice, the secret behind the mechanism of scapegoating and its role in society, as well as the effect of the Christian revelation on the latter.[1]

Girard starts with a simple but profound hypothesis that all human agency is based on desire and that desire is "mimetic," meaning that it is derived from someone else who signals the desirability of an object or good. Thus, all desire (not to be confused with appetites) is fundamentally social, not freestanding or autonomous as we tend to believe. If desire is mimetic, it does not simply seek out another desire to mirror it in solitude but imitates primordially to discover what it desires in the first

---

1. The following references offer summaries and synopses of Girard's work: Fleming, *René Girard*; Kirwan, *Discovering Girard*; Schwager, *Must There Be Scapegoats?*; Hamerton-Kelly, "Religion and the Thought of René Girard"; Williams, *Bible, Violence, and the Sacred*; Bailie, *Violence Unveiled*; Bartlett, *Cross Purposes*, 27–37; Mack, "Innocent Transgressor." Most articles in Wallace and Smith's *Curing Violence* contain brief summaries of Girard's theory. For a comprehensive exposition see Palaver, *René Girard's Mimetic Theory*.

place. Hence, desire as such remains always prereflective as it discovers itself in the world.

Girard demonstrates that desire arises when a model suggests in subtle gestures the desirability of an object or good so that we begin to admire it and then seek to possess it by imitating the model's desire for it. With this paradigm-shifting move, Girard calls us to completely rethink our understanding of the nature of desire, because most of us proceed on the assumption that what we desire is somehow grounded in our individual identity and core values or, as Freud has proposed, driven by sexual motivation.

Evidence for the validity of *mimetic theory* is manifold,[2] and its explanatory power has shed new light on some of the most agonizing problems of our contemporary world—for instance, the connection between secular modernity and religious terrorism, the entrenched and almost ritualized conflicts in the Middle East, and the contagious nature of violence and its role in culture and religion. It is therefore not surprising that some scholars have claimed that Girard has completely modified the intellectual landscape in many diverse disciplines such as the history of religion, philosophy, ethnology, psychology, literary criticism, economics, and theology.[3]

In the next section, readers not familiar with Girard's work will find a typical example of how mimetic desire plays out in human experience together with the destabilizing, even destructive effects the vortex of an imitative spiral generates.

According to Girard, we do not know what we desire until others point out what is desirable or worthwhile. In short, we take their desire as modeling our own. This simply means that our desire is *intersubjective*; it depends on the other, and it is this interdependence between the individual and the model that enables us to be moved by desires and passions in the first place.

To see more clearly how these reflexes work, let's go into a nursery. Imagine the scene. A child sits dreamily alone among assorted toys,

---

2. Today, "mimetic theory" as a general theory of culture has gained a life of its own. In various institutions of learning around the world, religious and literary scholars, psychologists, theologians, economists, lawyers, and others have begun to explore Girard's insights, especially the relationship between violence, culture, and religion. The contributions of the theological faculty at the University of Innsbruck (Austria) deserve special mention here.

3. Dumouchel, ed., *Violence and Truth*, 23.

showing only a casual interest in one of them. At this moment, another child enters the room. He sees the first child and in her vicinity a dozen or so toys. Now comes the moment when the second child is about to choose a toy. If you ask any parent which toy the second child is likely to find most interesting, they will invariably reply "the one the first child seems interested in," even though that interest is merely casual. What will happen next? As soon as the second child shows an interest in *that* toy, the desire of the first child for the same toy intensifies. She reaches for it a touch more energetically than before, clutching it more tightly. These gestures only heighten the desire of the second child. He stretches out his hand to take hold of it. This sets off an even stronger possessive response in the first child, who now raises her voice: "This is my toy!" Before long, an all-out squabble ensues over whose toy it is, although many others are readily accessible.

The progression of this conflict is quite predictable. When the acquisitive energies of two desiring subjects converge on the same object, they are prone to become rivals. At first mildly so, but not for long since the model (first child) also begins to imitate the desire of the imitating child, who in turn feels greater attraction for the toy with every acquisitive move by the model. Thus, the conflicted desires bounce back and forth, rising seesaw-like in intensity as each of the rivals reciprocates the acquisitive gestures of the other, albeit each time with greater force. The more model and imitator become mirror images of each other, the more they become locked into an insoluble "mimetic crisis." At that stage, the original object of desire is forgotten. The rivals no longer seek possession of a physical object but rather a metaphysical reality, namely, the special something the other seems to possess—the power to antagonize that now threatens their very existence.

Mimetic desire can be observed everywhere, not just in nurseries, and to a greater or lesser degree most of us have experienced its destabilizing, even destructive effects. The outworking of mimesis occurs in all kinds of social constructs and relations: in marriages, in board meetings, in local party politics, in the international arms race, and in the deadliest form of all, armed conflict.

The primary evidence for Girard's discovery of mimetic desire comes from the great literature he has studied to explain human behavior and the underlying "interdividual psychology." One of Girard's followers, Jean-Michel Oughourlian, professor of clinical psychology at the University of Paris, explains:

> What is this force that, from the beginning of life, draws the child into reproducing what an adult says or does? This force of attraction, interest, and attention is so much a part of the fabric of humanity that it is taken for granted. A young child has no power to resist that attraction. To feel such attraction is the child's very nature. . . . Similarly, if this remarkable force that attracts human beings to one another, that unites them, that enables children to model themselves on adults . . . did not exist, there would be no humanity. . . . This force, as fundamental for psychology as gravity is for physics, is what, following Girard, I call *mimesis*.[4]

What Oughourlian asserts here is the idea that mimesis belongs, like gravitation, to the fundamental realities of the universe and that without *universal mimesis* humans could not exist. Based on this premise, we must therefore see in Girard's discovery far more than a fancy explanation for the behavior displayed in the nursery scene. Indeed, what Girard has accomplished is the development of an incredibly rich theory of human culture and religion that looks at human nature as it is, as the neurosciences have begun to acknowledge. I have elaborated on the theory here primarily because it is time for the church and for individual Christians to take note, for any serious attempt to deal with our deepest nature can no longer ignore Girard's theory. After all, more than any other explanation, mimetic theory seems to illuminate our violent propensities and their contagious nature as well as ways of escaping their destructive spiral.[5]

After years of regarding imitation as irrelevant, experimental scientists in the social and cognitive sciences have at last begun to take it seriously to the point that imitation has now become a popular research topic. Although some still avoid the key element of Girard's theory—the conflict and drama caused by mimetic rivalry—mainstream cognitive science now recognizes that imitation is a pervasive human phenomenon.

---

4. Oughourlian, "From Universal Mimesis."

5. While one may argue that Christian references to humanity's "sinful" and "fallen" nature lack Girard's explanatory force, it is worth noting that the sin of concupiscence may be regarded as the biblical parallel of acquisitive mimesis as the garden scene of Genesis 3 reveals. What comes into view is the suggestive power of mimetic desire mediated by a proximate model (the serpent). As the mimetic exchange between the woman and the serpent progresses, desire magnifies, tipping the balance, and the woman's existential situation with all its wide open potentialities collapses in an irreversible actuality. Since then, according to the biblical story, this pattern inflects everything we humans do and say, even our perception of God, a bondage from which God seeks to save us.

Two discoveries in cognitive neuroscience have greatly helped our understanding of mimesis: the discovery by Andrew Meltzoff that infants at a very early age can read the intentions of adults even before their capacity for symbolic thinking has developed, and the discovery of so-called mirror neurons in research done under Vittorio Gallese at the University of Parma. The latter confirmed an earlier intuition that our mimetic propensities may be supported by some form of neurocognitive wiring that simulates in an observer's brain the current actions and feelings of others, suggesting that "empathy with others begins at a much lower level of embodied activity" than previously assumed.[6]

Before paying attention to the origin of culture from a Girardian point of view, we must address one crucial detail in the intersubjective makeup of mimesis. It concerns the interplay between the imitator and a model when the model belongs to a distant social domain. Picture an undergraduate student privileged to study under a famous professor, say a Nobel laureate. He admires his professor, is awed by his role model's ability to generate and synthesize complex knowledge, and is eager to learn under his tutelage. Aware of the great distance in stature that separates him as a student from his model, he learns from him by imitation as best as he can, deliberately and unconsciously. Although the student seeks his professor's approval, he will not compete with him. According to Girard, this kind of mimesis is "externally" mediated; it is a "peaceful" form of mimesis that does not lead to rivalry, also called "mimesis without an obstacle." In this good mimesis, the desire, although still acquisitive, is *nonrivalistic* or free of envy. However, the same desire can function in a different modality, as the sequel shows.

Fast-forward 20 years. Now the former undergraduate has become a well-known specialist in his field. He has published numerous books and articles and savors the prospect of obtaining a coveted professorship. But he is not the only candidate. A very respected senior colleague, with whom he is otherwise on good terms, has his eye on the same position. Since the two candidates occupy the same social domain, the possibility that mimetic rivalry will enter their relationship is far higher than in the previous scenario, with the potential to cause considerable strain. Girard speaks in this instance of "internally" mediated mimesis. To be sure, mimetic desire works in both cases, yet, depending on the social distance between the model and the desiring subject, the relational effect can be

---

6. Garrels, "Mimesis and Science," quote on 242.

quite different. This subtle distinction will occupy us again later in the chapter when we consider how to escape from bad mimesis.

As we continue to explore what mimesis reveals about us as human beings, we will pay special attention to its role in the origin of culture, because in order to survive early human groups needed to find ways of containing the destabilizing effects of mimetic rivalry. But first a word about how we moved from the prehuman state to human consciousness, or "the moment of hominization," as Girard calls it.

How We Became Human

What Girard's theory highlights is that *Homo sapiens* is gifted more than any other animal with the remarkable faculty for imitation. This faculty is grounded in a boundless capacity for observation and for reproducing what has been observed in action and attitude. As paradoxical as it may sound, this capacity is the basis for all learning and culture, yet also, at least potentially, for unstoppable violence. The latter is not, to begin with, the result of innate aggression or of a particularly combative nature but a manifestation of imitation of the kind we saw earlier in the nursery scene. In other words, conflicts are not always the result of scarcity as some economists have proposed. As we have seen, no matter how many "toys" are around, mimetic desire will always want the one toy that happens to be in the hands of the other.

When we surveyed the rise of human consciousness in chapter 6, we left unexplored how representational or symbolic expressions first emerged from a prehuman state. To explore this development here, we must take an *a posteriori* approach and reconstruct it from what we know now, because the moment when this happened is no longer accessible. Armed with Girard's insights into universal mimesis, we want to understand how the crucial moment came about when the instinctive responses of a prehuman group were breached to give way to a collective consciousness. Or, in another way, we want to speculate on the basis of Girard's theory what it was in their collective experience that generated the special something that in the future would enable the emergence of a sustainable social cohesion despite the destabilizing, even destructive effects of mimesis. After all, more mimetic than their evolutionary cousins, the great apes, these early human groups were especially open to *acquisitive* or *possessive* gestures and to their violent consequences. As these would have tended to overwhelm their instinctive controls, these groups would have been exceedingly vulnerable to reciprocal annihilation. Without the emergence of a mechanism with the power to override the mimetic descent into self-destruction, our early ancestors would not have survived. It is precisely the emergence of such a mechanism that Girard identifies as the point of "hominization" or the moment we crossed the threshold from the prehuman to the human. This moment is also the *founding event of human consciousness*, the point in evolutionary history when human brains for the first time became aware of meaning as they underwent a symbolic experience not generated by a purely instinctive reaction. Let us follow Girard in slow motion.

Assume a scene in which a typical mimetic conflict between two individuals arises within a group of hominids; assume further that before long the squabble spills over into violence. Because of its contagious nature, mimetic violence draws more and more bystanders into the fray, pitting more and more members of the group against each other. With rising intensity, the fracas deteriorates into an all-against-all frenzy that threatens the life of the group. In the heat of the battle, the object that provoked this "war" is forgotten, as the rivals keep imitating each other, thereby erasing all differences between them. As the mimetic crisis progresses, all violence gets directed to combat the very force that each feels is coming against him, a force that emanated with uncanny potency from all the members of the group. At the height of the crisis, as spontaneously as it arose, the volatile mimetic animus of the group flips from an all-against-all to an all-against-one conflict. The mutually destructive energy now unites the group against one arbitrary victim on whom the group unanimously vents the full force of their collective violence, and who cannot retaliate. The result? Mob violence kills one of their own kind. By slaying the victim, it is expelled or sacrificed. Instantly, hostilities cease, the vicious circle of mutually destructive rage is breached, and the war is over. In stunned silence and peaceful unanimity, they stare at the corpse, the victim of their collective wrath. As unnerving as the fury that preceded it is the sudden peace and unity that followed the storm. Transfixed by the presence of the dead companion and the sudden mind-numbing reversal from war to peace, their collective consciousness would have been flooded with terrifying dread before this strange power that first appeared as a wave of destructive violence and now comes to their rescue. Girard calls this power by a name: the "*primitive sacred*."

In this solemn moment, Girard theorizes, a faint (mis)recognition dawns on them that attributes whatever they have experienced to the dead victim (they have no knowledge of the hidden mimetic mechanism). Girard calls it the secret kept "hidden from the foundation of the world" (Matt 13:35). As a result, the victim becomes the monster that caused their trouble, on the one hand, and their rescuer, on the other. This, in Girard's conception, is the moment of "hominization," the "scene of origin," when during an outbreak of unstoppable violence in an archaic group a countermechanism emerged that could pacify the blind fury after mimesis spiraled out of control, threatening the group's life. Because the group is preserved *by the expulsion of the victim* in situations like this, Girard calls it the "victimage" or "scapegoat mechanism," whose spontaneous emergence stands at the beginning of archaic religion and of human culture.

After all, had it not been for the appearance of this mechanism, archaic groups would have collapsed when the evolution of the prehuman brain had reached the point where mimetic urges outran the dominance pattern that prevails in the animal world. Groups that discovered the mechanism survived and moved on as the mechanism itself began its own evolutionary path toward new and more complex cultural forms; those groups that did not became extinct.

Since this process also gave rise to acts of attention not grounded in instinctive reactions (e.g., when mimetic violence ceased), it lies at the root of our symbolic intelligence, which we explored in chapter 6, along with diverse individual and group identities. In his religious and cultural elaborations, Girard is no longer concerned with the first *definitive act of humanity*, but with the scapegoat mechanism itself and the mythical concealment of its violence, which constitute the foundations of religion and culture, as we will see in the next section.

On Sacrifice, Scapegoats, and Victims

If you lived anywhere in the world in, say, 3000 BCE, you would have believed in the gods and participated in the sacrificial cult of your village, city, or people group. What you believed about the gods (your theology) was not very important. The key to your welfare and that of your family and community lay in the ritual precision with which the cultic actions, usually sacrifices, were performed. If done correctly, the sacrifice was accepted and the worshipper(s) would be blessed. The larger and more costly the sacrifice, the better. Many ancient cultures practiced human sacrifice to please the gods or to appease their wrath. In all cultures, natural disasters like earthquakes, heavy rainstorms or droughts, or a plague of locusts were understood as signs of divine anger, requiring costly sacrifices. Special occasions like the death of a king or the dedication of a temple or religious monument (e.g., Stonehenge) were accompanied by human sacrifice. This feature of archaic religion has marked human culture for millennia, from Scandinavia to China, from Mesopotamia to Mexico, until quite recently. In 1777, on one of his visits to the Polynesian islands, Captain James Cook (1728–1778) was invited to witness a human sacrifice, which he meticulously recorded in his diary. The islanders were seeking assistance from the gods (whose favor could allegedly be secured by the offering of a fresh corpse) in a war looming with a neighboring island.[7]

The religion of ancient Israel too demanded various sacrifices and offerings, as described in Leviticus, the third book in the Old Testament. While bloodless sacrifices were included, animal sacrifices were most important. There was first the "whole burned offering" (*olah*); this standard offering was a voluntary sacrifice to which the cult attributed a high degree of solemnity and sacredness. An entire animal, except for its hide, was burned on the altar (Lev 1:1–17). The "meal offering" (*minchah*) consisted of flour, oil, salt, and frankincense; it was partly burned on the altar and partly given to the priests to eat (Lev 2:1–16). The "sacrifice of well-being" or "peace offering" (*zevach sh'lamim*) was a voluntary animal sacrifice, male or female, from one's herd; certain details had to be observed in terms of how to burn the fat as well as the kidneys and other organs (Lev 3:1–17). The "sin offering" (*chatat*) was an obligatory animal sacrifice to atone for unintentional sins. It prescribed special treatment of the blood, which distinguished it from other sacrifices (Lev 4:1–5:13). The "penalty" or "guilt offering" (*asham*) was an obligatory sacrifice of a ram; it was required for misappropriating property (Lev 5:1–26).

As at least two texts show, the Jewish scriptures point to an awareness in Israel of human sacrifice: Abraham's near sacrifice of Isaac and the fate of Jephthah's daughter (Gen 22:1–15 and Judg 11:1–34). Other sources attest that the king of Moab gave his firstborn son as a burned offering, while the Amorites are reported to have sacrificed children to

---

7. Bailie, *Violence Unveiled*, 67–85.

the Canaanite deity Moloch. The Old Testament prohibits human sacrifice, and child sacrifice is considered especially abominable (Lev 18:21, 20:3; Deut 12:30–31, 18:10; Ps 106:37), yet the fact that these prohibitions were necessary seems to suggest that they played a role in the religion of early Israel. While sacrifices ceased in Judaism after the destruction of the Jerusalem temple in 70 CE, the sacrificial mindset persisted in the tradition, as well as in Christianity, although there in a highly transmuted sense. More on that later.

Globally speaking, the sacrificial mentality of archaic religion that demands corpses is still alive even in the twenty-first century. According to UN reports from Nigeria, Uganda, Swaziland, Liberia, Tanzania, Namibia, and Mali, the ritual killing of children and adults in human sacrifices is still in vogue in Sub-Saharan Africa.[8]

Structurally, the practice of sacrifice involves a bargain or exchange: one gives something costly to the gods and the gods are asked to do something in return; the bigger the problem, the more costly the sacrifice ought to be. Child sacrifice extends another religious idea that the more costly the sacrifice, the greater the piety of the one who offers it. Repetition of such gruesome events and their mythical obfuscations over the millennia has so preconditioned human consciousness that the underlying violence has disappeared from view, and we owe it to the genius of René Girard to have unearthed it. Because of its significance, it deserves to be explored more fully.

As noted earlier, Girard associates the "scene of origin" with the horror of an outbreak of chaotic violence *within* an archaic group that spills over into their social space. The crisis fills the group with dread and brings about the experience of an encounter with the sacred (see chapter 7). This notion has important implications for Girard's interpretation of the nature of religion, especially of sacrifice.

He argues that in the peak moment of the crisis the group cannot see that the victim was arbitrarily chosen and could not have been responsible for their woes. Such clarity would have left the conflict unresolved, but since it was resolved in the all-against-one act of victimization, the act becomes sacralized. Eventually, the expelled victim is seen as the visitation of a god who brought chaos followed by order and peace, so that what the victim represented was to be held in awe. Since such

---

8. See Dateline Current Affairs, www.sbs.com.au/news/dateline/ why-child-sacrifice-is-on-the-rise-in-uganda; and, https://en.wikipedia.org/wiki/Child_sacrifice_in_Uganda.

crises repeated themselves, always with the same result, the group began to learn the pattern and could in the future maintain its social cohesion by ritualizing the original scene. After the evolution of language, groups would begin to tell each other stories about this founding event as part of their group, concealing the murder behind the veil of sacred violence. With time, groups developed their own traditions, including prohibitions against behaviors that would lead to outbreaks of violence.

Key to understanding Girard's anthropology is the conception that the cohesion of archaic sociality is achieved by ritually reenacting the original scene based on stories about its origin. Yet, these stories are not to be mistaken for historical reports. They are myths whose nature is to attribute responsibility for the foundational murder to the god or gods that appear in the myths. For all their prehistoric patina, these insights are highly relevant today.

Fundamentally, then, the violence that belongs to acquisitive mimesis has rendered all human social life very fragile from the beginning. Because mimetic crises were part of the earliest human society and were resolved by sacrifice, Girard argues that culture and religion were attempts to lessen this fragility, thus making life more secure. Consequently, sacrifice is the primordial institution of human culture; it includes the scapegoat mechanism that evolved with sacrifice. Both are rooted in a mimeticism that is far more intense in humans than in other creatures.[9]

But let's go back to the uncanny force that originally united the group during the crisis and so precipitated the anonymous killing, reconciling the participants in their belief that sacred violence killed the victim. By assigning blame for the trouble to the slain victim, the destructive momentum of mimetic violence also dies with it, being transposed into a protective influence that now preserves the life of the group and its future. In Girard's conception, then, sacrifice was a way of holding the impulse of reciprocal violence (and vengeance) in check in the guise of religious violence.[10] Sacrifice so understood is an anthropological datum as well as a religious one.

One more point needs to be made. Sacrificial practices have been repeated many times over thousands of years without losing their effect, as the victimage mechanism has provided a workable tool to hold society together while leaving the ambiguous role of mimesis in place. Later in

---

9. Girard, *Sacrifice*, 62.
10. Girard, *Violence and the Sacred*, 12–14.

the chapter, I explain why mimesis needed to be preserved, and why in society scapegoats are indispensable. Understandably, Girard rejects the idea that it is natural for human beings to live in peace with each other.

If today we find it more difficult to believe that the victim is guilty, Girard attributes this weakening of the victim mechanism to the demystifying effect of the Christian revelation. At the same time, institutional Christianity must bear its own share of responsibility for the present state of the world. As Anthony Bartlett has shown, for 2,000 years the story of the Crucified, especially interpretations of Jesus' death, has led to an astonishingly violent history of the Christian religion.[11] A brief survey later in this chapter of the God language that developed in their wake will shed some light on this bewildering reality. But first, a deeper look into the issue of how our individual identities are constituted under conditions of universal mimesis.[12]

The Mimetic Constitution of the Human Person— Homo Imitans

In the last section I emphasized how an archaic group that, moments earlier, had been in the grip of a mimetic crisis became pacified after the frenzy and focused their attention on the corpse of the victim. This attention emerged not from instinctive reactions but from within the liminal

---

11. Bartlett, *Cross Purposes*, 1.

12. While Girard presents us with a compelling hypothesis, it does remain controversial. His critics are mostly concerned with the validity of his assumptions and the radicalism of his claims. German Catholic theologian Eugen Biser, for instance, rejects mimetic theory outright as an "absurd thesis" (Biser, "Die Stimme der Antigone"), while Herzog takes Girard to task for drawing all of human history into a single event type, the "Ur-sacrifice" (Herzog, "Religionstheorie"). Similarly, Finlan charges Girard with reductionism over his single-issue anthropology as the foundation of religion and culture. On the other hand, Finlan acknowledges the timeliness and necessity of Girard's theory and the astuteness of Girardian exegetes at analyzing both violent stories in the Old Testament and the divine violence in other biblical authors (see Finlan, *Problems with Atonement*, 91–93). Anthony Bartlett, who builds his entire work on Girard's theory of universal mimesis, also criticizes his exclusive focus on a "genealogy of violence" without considering a "genealogy of compassion" (Bartlett, *Cross Purposes*, 40). Feminist critics have argued that Girard's theory does not adequately account for women as protagonists and victims of violence (Kirk-Duggan, "Gender, Violence, and Transformation," 244). Girard, on the other hand, is at pains to show in his responses that his critics tend to overlook the universality of mimesis, and not violence, as the primary factor in both culture and religion, and that his "originary scene" predates by tens of thousands of years the emergence of religions studied by religious historians.

experience into which their collective consciousness was plunged. In this instance of awe mimesis was inoperative, suspended: no desirable object in play, no acquisitive gesture to imitate, no alpha male to be envied, for envy itself seems to have been eliminated. In this hush, even normal appetites for food and sex had fled their consciousness. Everything had become untouchable. What did fill their collective consciousness, what commanded their first noninstinctive attention, was the corpse of the slain victim!

Not that they were unfamiliar with the fact of death or had not seen the dead bodies of companions and prey. What mattered was that this corpse and this death were different. Both had arisen inexplicably in the fury that had engulfed their own ranks, threatening to destroy them. What now overwhelmed and arrested their emotional response was the rapid reversal from frenzy to unanimous calm and sacred awe in the wake of the killing.

If this imaginative sketch comes close to what may have happened in archaic communities everywhere, it lends plausibility to Girard's idea that the experience of the collective murder combined with the moment of sacred awe would have inflected the development of all human culture. Also, that such highly emotional experiences should have impressed themselves deeply on the evolving brain of our early ancestors is a conclusion that modern neuroscience would support. Given the meaning-making capacity of the brain even in its early stages, it would most likely have woven the three elements of the experience—the victim, the scapegoat mechanism, and the presence of sacred awe—into one, thereby syncing all subsequent religious and cultural institutions to the knowledge of a victim and its violent death.

One of Girard's most astute theological and cultural interpreters, James Alison, puts it this way: "Every institution of human culture is shot through with violent mimesis. All human sociality is born thanks to the victim, and particularly, to ignorance of the victim(s) that gave it birth. Human language and thought are already utterly inflected by this ur-violence from the conception."[13] Moreover, we must remind ourselves that every infant is inescapably cast into a mimetic matrix from conception (which is itself grounded in a highly mimetic act). According to developmental psychology, the first memories of neonates—their cooing sounds, gestures, and facial expressions—are all mimetically constituted, just as

---

13. Alison, *Joy of Being Wrong*, 16.

a child's language is in later stages of growth.[14] Even preverbal children show early forms of the ability to coordinate social and affective responses and the ability to communicate intentions, even share in pretend games.

If mimesis is universal today, the same must have been true for all preceding generations, and it would have worked the same way in earlier stages of human development everywhere. Because imitation is the powerful learning tool that it is, it enabled every new generation to learn from the previous one, avoiding the errors of their predecessors.[15] In other words, we are much more *Homo imitans*, "man the imitator," than we are *Homo sapiens*, "man the knower."

What's more, imitation requires that there is someone else prior to the infant, like the primary caregiver. If mimesis is the imitation of the other's desires, then these too are prior to the infant, so that everything that becomes the *I* or the *me* through imitation is based on something previous. In short, the rather impressionable identity that is forming is "radically dependent on the desires of others." Thus, the unavoidable working of universal mimesis renders our own sense of originality illusory as far as our desires go, because they do not have *me* as the source but "me-imitating-the-desires-of-others."[16]

This misrecognition is not a mistake in the ordinary sense. We simply could not have recognized our dependence on the desires of others as something separate from our consciousness. While mimesis generates any number of different and differentiated *me*s that theoretically could live peaceably alongside each other, "so long as they relate as models, not rivals," such a state is in practice beyond the realm of possibility. For all infants born into the world, no matter how loving or nonrivalistic their upbringing, are predisposed toward rivalry simply because the human culture that shaped their models and whose imitation constituted them as persons is rivalistic at root.[17] In other words, the signals and suggestions the infant receives are all weighted toward rivalistic mimesis and the mechanism of victimage, for without it no social and personal identity can exist. What's more, *stability of identity crucially depends on the principle of differentiation or expulsion.* As a result, every infant finds itself from the beginning in an ambiguous world of rivalry where desiring is

14. See Meltzoff and Moore, "Imitation."
15. Garrels, "Human Imitation," esp. 21.
16. Alison, *Joy of Being Wrong*, 16–17.
17. Alison, *Joy of Being Wrong*, 17.

competitive, first feebly while overpowering adults are the models and for a long time as an inadequate learner. We must also consider what we saw earlier—that eventually the objects of desire become irrelevant. In a quarrel, what the other *is* is more important than what the other *has*. The quarrel, then, is not over the possession of this or that toy, but over identity, and since identity can only be established by exclusion, we are back in the hands of the victimage mechanism. It serves, in the final analysis, to (re)establish difference and arises with special force in situations "where everyone resembles everyone else."[18]

As a result, we humans are desperately seeking to establish distinctions without realizing that the *me* that so seeks to establish itself is a phantom based on other-aroused desires mediated to us by generations of models. If these other-aroused desires that masquerade as *me* and covet distinctions represent an inevitable mimetic occasion, it follows that this phantom *me* must be a *me-against-others*. And it is this *self* that lies behind our feelings of moral indignation when we are put upon, behind the sense of duty to stand up for ourselves, behind high-pitched political campaigns, and more. This is the *self* we cultivate, that receives all our energy, and only when we see with renewed eyes will we realize that we are spending this energy "on maintaining the illusion of an independent, autonomous self—as opposed to . . . [the Model] that comes to me as a gift from the Father of Lights," as Sebastian Moore reminds us.[19]

The same dynamics that establish individual identities are at work in the formation of group or institutional identity. Here too distinctions are established in an over-and-against fashion, often carried to an extreme, as the bloody examples of secular and church history show. Regardless, whether the identity is tribal, ethnic, national, imperial, ecclesial, or religious, forming or maintaining it requires a sacrificial solution of one kind or another. After all, identities mean an entity vis-à-vis others and that calls for an us-and-them awareness, which is created by setting clear boundary markers, often reinforced by qualifying belief structures (creeds, manifestos). When these are in place, those who do not qualify as insiders can be spotted and treated with intolerance at best, if not with persecution.

Even the framing of Christian orthodoxy was not exempt from this pattern. Think of the intense rivalry between Arianism and Trinitarianism

---

18. Girard, *Battling to the End*, ix.
19. Moore, "Foreword," viii.

in the fourth century, of the inhuman severity with which the Catholic Church sought to root out heresy in Europe and the Americas, of the crusaders who massacred Jews in Europe during the eleventh century, of Anglican brutality toward Catholics during the English Reformation,[20] of Martin Luther's violent invective against the Jews and its continuing influence on modern anti-Semitism.[21] Thousands, if not millions, were murdered based on the implacable logic of the scapegoat that pronounced the victims guilty as charged. When we recognize this for what it is, we are confronted again with the collective violence that lies shrouded in myth and seemingly forgotten, yet inflecting everything we think and do, including the way we conceive of God.

In the next section we will trace the development of God language based on the way Christian theology has interpreted the death of Jesus. The aim of the following brief survey is to compare conceptions of God that developed in Western Christianity with the image Jesus projected when he spoke of God in vocative language as *Abba*. Suggesting tenderness and compassion, the word makes us think of a human father bending over the crib of his newborn baby. It is an image that speaks to us of the intimacy Jesus himself experienced in communion with his Father, reinforcing the portrait of God Jesus composed in his parables (e.g., the Good Samaritan and the Prodigal Son, etc.), seeking to pass it on to his disciples. That this portrait confers a special interpretive touch on Jesus' exclusive claim in Matthew's Gospel cannot be doubted: "All things have been handed over to me by my Father; and no one knows the Son except the Father, and no one knows the Father except the Son and anyone to whom the Son chooses to reveal him" (Matt 11:27).

---

20. Head, "Herodian Aspects."

21. Luther, *On the Jews and Their Lies*, 268–71. While one of Luther's biographers observed that Luther's hostility toward the Jews should not be regarded as "psychological" or "pathological" hatred, nor as a politically motivated ideology, Luther "demanded that measures provided in the laws against heretics be employed to expel the Jews—similarly to their use against the Anabaptists—because, in view of the Jewish polemics against Christ, he saw no possibilities for religious coexistence." The long-term effect was that "his misguided agitation had the evil result that Luther fatefully became one of the 'church fathers' of anti-Semitism, thus providing material for the modern hatred of the Jews, cloaking it with the authority of the Reformer" (Brecht, *Martin Luther*, 350–51).

The Crisis of God as Abba

God is of central importance in the way Christians think and live their lives. As indicated above, God language can be ambiguous, so that the meaning of the word *God* always needs clarification. Let me begin with Jesus' emphasis in proclaiming the good news of the "kingdom" (we find nearly 350 references in the four Gospels and the book of Acts). Yet, in his God language Jesus did not speak in monarchical terms of sword and crown. Rather, he called God *Abba*, adopting an intimate familial term. When challenged by Pontius Pilate about his alleged kingship, he deflected the question, saying that his kingdom "was not of this world" (John 18:36). He suggested that the kingdom he proclaimed—the *basilea theou*—had foundations and prosecuted an agenda that differed radically from all the kingdoms of this world. Those for whom this kingdom resonated displayed different values than the members of the prevailing social order did. Aware of their spiritual poverty, they depended on *Abba*'s fatherly benevolence, craved his righteousness, and were willing to suffer persecution for his sake. When mistreated, they turned the other cheek; when forced to, they walked one mile, then walked another; and, in their commitment to peacemaking, they followed Jesus' nonviolent model to the point of laying down their life for their friends. Imitating Jesus (in whom they saw his *Abba*), they abhorred domination, coercion, rivalry, and violence in any form for any reason, for their supreme value was love.

When Jesus spoke of the kingdom of heaven, he was not referring to an otherworldly existence, the kingdom's otherworldly origin notwithstanding. The *basilea* is to operate in this world ("*on earth* as it is in heaven"), whereby *Abba* participates in the subjective experience of those in communion with him, working with them, while trinitarian love draws the creation toward its future consummation on a cosmic scale. On all accounts, explicitly and implicitly, *Abba*'s agenda is for the living to have more life, more abundantly even now (John 10:10), revealing the values this agenda seeks to actualize in history, presumably with important implications for historical Christianity. After all, Jesus revealed that with the word *God* he meant none other than his nonviolent *Abba*. On this account alone we are faced with the crucial question of whether historical Christianity has been true to that conception and whether it is essential to recover that revelation in our time.

To that end, I will review aspects of five influential theologies that have shaped Western Christianity in the postapostolic era, highlighting two crucial points. First, in developing its God language, historical Christianity has followed cultural-institutional models with deadly consequences that critically obscure the portrait Jesus sought to instill in his followers. Second, the resultant loss in revelatory continuity might be recovered, so that in a world of selfish and violent discontinuity a Christian vision might direct our gaze in hope toward the transformative power of love. Continuing to rely on René Girard's mimetic theory, as modified by Anthony Bartlett's groundbreaking application, the result is a reconception of God as having nothing to do with violence but only with abyssal compassion and its transformative effect, which seeks us out in our anxiety and alienation.[22]

By the twelfth century, as we will see, theological developments in the Christian West had reinterpreted the death of Christ in a manner that rendered proximate and deferred violence absolute in the person of God. What's more, this conception would become Christianity's dominant Western account even to this day. Let's look at these developments more closely.

Ambrose (340–397)

Ambrose, bishop of Milan, was one of the most revered and influential figures of the fourth-century church. Like other Latin Church fathers, he was trained in Roman law and developed the idea that the death of Jesus Christ was a transaction in civil justice (i.e., a mimetic exchange), whereby our ancestor Adam incurred a debt to the devil that was then passed on to his offspring with accumulating interest. Christ by his death wiped out both. Ambrose understood the incarnation as a reciprocal fraud that countered the devil's earlier fraud (in the garden), but the second fraud, the resurrection, was holy. He also saw Christ's death as penal substitution in that the debt to the devil was also punishment administered by the devil, although God was the final arbiter. As Bartlett points out, this theme arises inevitably out of the logic of mimetic exchange; what is deferred violence in the case of the devil becomes appeasement or expiation when God, the final judge, is in view.[23]

---

22. Bartlett, *Cross Purposes*.
23. Bartlett, *Cross Purposes*, 71n74.

Augustine (354–430)

Augustine, bishop of Hippo, developed the idea of debt to the devil further. In *De Trinitate*, he wrote:

> in this redemption the blood of Christ was as it were the price given for us (but the devil on receiving it was not enriched but bound), in order that we might be loosed from its chains ... and that he might not ... draw with himself to the ruin of the second and eternal death, any one of those whom Christ, free from all debt, had redeemed by pouring out his own blood without being obliged to do so.[24]

According to Bartlett, Augustine saw in the "chains" a "real ontological indebtedness in respect of a personified Devil," which sets God and the devil up in a scheme of mimetic rivalry.[25] The frequency with which Augustine referred to this exchange seems to indicate how settled it was in his mind. Moreover, since this scheme was conceived to exist within God's overarching judgment, the devil was not an independent agent, but "God's executioner of God's supreme judicial decisions."[26] In other words, penal substitution and the transaction with the devil merge into one. Bartlett writes:

> More significantly [this point] is subordinate to Augustine's final view of God, an arbitrary dispenser of mercy in a sea of condemnation. It does not cross his mind that the elaborate descriptions of the mimetic set-off or exchange between Christ and the Devil lose all redemptive meaning if their positive effect is a foregone conclusion in the mind of God. Here the Crucified is a passive function of a remote and indifferent will, and ultimately no more than a kind of window-dressing of mercy while inside the house a throne of violence rules.[27]

The conclusion? Augustine's elaborate scheme of a mimetic exchange between God and the devil loses its interpretive value because behind it there is the inscrutable will of God and his ultimate violence, which Christ averts for the elect. According to Bartlett, if in Augustine's scheme "abyssal love [were] substituted for transcendent will," it would amount

---

24. Quoted in Bartlett, *Cross Purposes*, 71n75.
25. Bartlett, *Cross Purposes*, 72.
26. Bartlett, *Cross Purposes*, 73.
27. Bartlett, *Cross* Purposes, 73.

to "an unconditional act of redemption."[28] As it is, it falls short despite its positive aspects. His theology influenced Christian thought for centuries. Martin Luther was an Augustinian and John Calvin's doctrinal system[29] goes back to the Augustinianism common to all Reformation theology.

Gregory the Great (c. 540–604)

Son of a Roman senator, trained in Roman law, Gregory the Great was the prefect of Rome at a young age, and he became the first monastic pope or bishop of Rome. Known for his administrative skills, missions, and prolific writings, he was "the primary popularizer of Augustine and chief mobilizer of the typical mindset of medieval faith."[30] For our purposes he is uniquely important on two counts. Known throughout the Middle Ages as the "father of Christian worship," he is still recognized for his teaching on the Eucharist and penance. As a missionary pope, he also converted the Anglo-Saxons and realigned the (Arian Christian) Franks, Lombards, and Visigoths with Christian orthodoxy.

By then, the relationship between God and humanity had standardized based on the Roman legal model, which Gregory continued to disseminate. This legal model also influenced his view of the Eucharist and penance. The body and the blood of Christ were present in the elements necessarily, derived from the cultic principle that the Eucharist was performed by a priest as a real repetition of Christ's sacrifice. However, there was a difference. While Christ's death on the cross availed for the sins of all the elect, its repetition was meritorious only for the participants (and those for whose benefit it was specifically offered). Eucharist and penance evinced the same effect; they deferred or preempted divine violence: "For either man himself by penance punishes sin in himself, or God taking vengeance on him smites him."[31] In other words, a century after Augustine the scheme of divine violence had become so deeply established that another place separate from hell was needed—purgatory—where the elect were to undergo whatever punishment was outstanding at the time of their death. This was an emphatic theme for Gregory.

---

28. Bartlett, *Cross* Purposes, 73.

29. The five points of Calvinism are: *Total depravity of man, Unconditional election, Limited atonement, Irresistible grace, Perseverance of the saints* (TULIP).

30. Bartlett, *Cross Purposes*, 75.

31. Bartlett, *Cross Purposes*, 74, citing Gregory, *Moralia* 9.54.

In cosmic terms, Gregory's world was turbulent, full of armed conflict between angels, demons, and saints. This cosmos was presided over by Christ, the supreme judge and warrior Savior, who had done battle with the devil and won. Bartlett puts it this way: "Almost at every turn this universe was posited on institutionalized mimesis, a constant daily practice that negotiated the possibility of human security in the face of eternal violence."[32] In an important footnote, however, he adds that Gregory's teaching, based on his monastic-contemplative background, was also moving on a different level to overcome conflictual mimesis through loving desire for God, which, when experienced as "absence," leads to compunction of the heart (*compuntio cordis*).[33]

The second reason Gregory is of special importance in this survey is that during his papacy the tribes of northern Europe began to embrace Christianity, bringing with them a warrior ethos that effectively reformulated Christianity. Bartlett writes: "Personal honor, fealty, and fierce folk religion extruded a vigorous new landscape over older features of imperial order and intellectual faith."[34] These new cultural forces consolidated over the next few hundred years, raising new theological questions.

Anselm of Canterbury (1033–1109)

Drawing on these developments, Anselm of Canterbury produced a new and dynamic account of God's relationship with humanity. A Benedictine monk and later archbishop of Canterbury, Anselm was greatly influenced by Augustine's conception of God. Yet, he was also the first to move beyond Augustine's legacy by "reaffirming the historical-cultural value of the Redeemer."[35]

His landmark work *Cur Deus Homo* (*Why God Became Man*), composed between 1094 and 1098, is without a doubt one of the most significant works on the Christian doctrine of atonement, shutting the door on older interpretations of Christ's death and pointing forward to the now-common default model known as the satisfaction theory of atonement. In the Western theological canon, this medieval treatise has

---

32. Bartlett, *Cross Purposes*, 75.
33. Bartlett, *Cross Purposes*, 75n88.
34. Bartlett, *Cross Purposes*, 75.
35. Bartlett, *Cross Purposes*, 73.

achieved almost sacred status. To understand its preeminence, let us begin by looking at the circumstances that gave birth to it.[36]

When *Cur Deus Homo* was composed, eleventh-century Europe was on the verge of an intellectual renaissance awakened by a flourishing feudal and monastic culture. Hence, Anselm's work must be seen as a defining, even celebratory juncture in cultural history that also reflected clear political and theological objectives. Toward the end of the book, his discussion partner in the dialogue, Boso, joyfully affirms, "In proving that God became man by necessity ... you convince both Jews and Pagans by the mere force of reason."[37]

What was it, then, that should have obliged God to take such a personal step? According to Bartlett, this "obligation" had much to do with the intellectual challenges the Christian world of the time was facing, including the monastic ideology. For one, Jews in Europe were increasing in number, and their intellectually more advanced rabbis, who strongly opposed Christian doctrines, outmatched most Christian thinkers of the time. Moreover, during the composition of *Cur Deus Homo*, Pope Urban II mobilized the First Crusade in 1096, an event that unleashed the first Christian mass murder of Jews. A Yale scholar and historian of medieval intellectual history, Jaroslav Pelican, has observed: "Presumably it was the expression of some such judgment [that Jews were in their midst] when the Crusaders, on their way to make war with the Muslim infidels in the Holy Land, interrupted their journey to massacre Jews in Europe."[38]

From this and other accounts we may conclude that medieval Christianity was in the grip of intense currents of mimetic rivalry that surged through and shaped the social, cultural, and theological world, including the arguments of Anselm's book. In such a climate, the charge that infidels were ridiculing Christian doctrines constituted a grave assault on God's justice and honor, for Anselm, offending his own deeply held sense of feudal honor. Because Anselm defined the relations between God and man in feudal terms, whereby God was the lord and humanity his vassals, diminishing God's honor was an outrage of cosmic proportions, as it threatened everything. As Bartlett puts it, "To threaten the Christian

---

36. Bartlett, *Cross Purposes*, 76.
37. Bartlett, *Cross Purposes*, 77.
38. Cited in Bartlett, *Cross Purposes*, 79n100.

God, the supreme Lord, was to threaten the capstone and cement of the universe."³⁹

Since honor is a classic mimetic category, avowals such as these reveal the mimetic mindset from which Anselm's satisfaction theory of atonement springs, best summed up in his words: "In the order of things, there is nothing less to be endured than that the creature should take away the honor of the Creator, and not restore what he has taken away." From Anselm's feudal perspective, the creature owes the Creator honor in the same way a serf owes honor to the lord of the manor. In that scheme, sin breaks the agreement because it dishonors God, for sin is fundamentally an indignity to God, tarnishing his honor. Since the lord of the manor cannot afford to let his servants show disrespect without upsetting the entire system, there must be satisfaction if that order is to be restored. In his attempt to fit God and humanity into the feudal order, Anselm raises the category of honor to the level of the absolute: "He who violates another's honor does not enough by merely rendering honor again, but must according to the extent of the injury done, make restoration in some way satisfactory to the person whom he has dishonored. . . . So then, everyone who sins ought to pay back the honor of which he has robbed God; and this is the satisfaction which every sinner owes to God."⁴⁰

Since no human being could offer adequate satisfaction to restore the divine order, God had to become man so that Christ's death recompenses and makes reparations for dishonor occasioned by human sin. At the same time, Christ's undeserved death earns a surplus of merit that suffices to pay off humanity's sin debt.⁴¹ But that's not all.

Certainly, the logic that Jesus pays the debt incurred by sinful humanity with his suffering and death has been deeply embedded in Christian consciousness since *Cur Deus Homo*. However, we still need to account for a fact that Anselm is quite clear about, namely, that the notion of satisfaction is a structured form of vengeance, either satisfaction or torment:

> It is impossible for God to lose his honor, for either the sinner pays the debt of his own accord, or, if he refuses, God takes it from him. For either man renders due submission to God of his own will, by avoiding sin or making payment, or God subjects

---

39. Bartlett, *Cross Purposes*, 80.
40. Anselm, *Cur Deus Homo*, I, 22, 231, cited in Bartlett, *Cross Purposes*, 83.
41. Rashdall, *Idea of Atonement*, 351.

> him to himself for torments, even against man's will, and thus shows that he is the lord of man, though man refuses to acknowledge it of his own accord. And here we must observe that as man in sinning takes away what belongs to God, so God in punishing gets in turn what belongs to man.[42]

Since the devil no longer features as the violence-inflicting agent in Anselm's scheme, the only source of transcendent violence by which satisfaction is achieved is God the Father. Christ's death then becomes a substitute for revenge, which opens another theme, the inner conflict between the justice of the Father and the mercy of the Son. Since penal substitution is the "logical realization of this opposition," it also exposes (but never resolves) "the terrifying sacrificial crisis" that unleashes the violence of both God and man against the innocent victim, permanently establishing the need for a scapegoat.[43] Bartlett calls it "the crisis of God." While Anselm preserved the divine unity and cosmic sovereignty of God, in doing so he established divine violence "formally and metaphysically,"[44] creating in *Cur Deus Homo* "a master text of divine violence."[45] It was precisely this metaphysical force that ensured the enduring success of Anselm's legacy in Christianity's doctrinal history.

Two hundred years later, Thomas Aquinas fully integrated "satisfaction" in the medieval doctrine of redemption: "As therefore Christ's passion provided adequate, and more than adequate satisfaction for humanity's sin and debt of punishment, his death was as it were the price by which we are freed from both obligations. Satisfaction offered for oneself or for another is called the price whereby one ransoms oneself or another from sin and from punishment."[46] From there the idea of satisfaction continued to influence Catholic theological reflection, as the *Catechism of the Catholic Church* makes clear: "Jesus atoned for our faults and made satisfaction for our sins to the Father."[47] This leaves unexplained the underlying problem of divine violence, which so glaringly conflicts with the teaching and conduct of Jesus, whose identity with the Father—whom he calls *Abba*—is central to the New Testament revelation, affirmed in over

---

42. Bartlett, *Cross Purposes*, 84.
43. Bartlett, *Cross Purposes*, 85.
44. Bartlett, *Cross Purposes*, 66.
45. Bartlett, *Cross Purposes*, 76.
46. Aquinas, *Summa Theologiae* 3a.48.2.
47. *Catechism of the Catholic Church*, para. 615.

100 references: "I and the Father are one," and "Whoever has seen me, has seen the Father" (John 10:30, 14:9).

Martin Luther (1483–1546)

When the Protestant Reformation adopted Anselm's doctrine 400 years later, the reformers drove the "crisis of God" to new levels. No longer concerned with divine honor, they went straight to penal substitution and its underlying rationale, divine wrath. Martin Luther's graphic and forceful exposition of Christ's substitutionary death went further than anything theologians had imagined before: Christ was made a curse for us, physically bearing in his body all sins of all humanity. By being physically identified with their sins, he became the worst offender: blasphemer . . . adulterer . . . murderer . . . deceiver . . . not in the sense of having committed these sins, as Luther wrote, but by "really and truly [offering] Himself to the Father for eternal punishment on our behalf." Luther continued: "His human nature behaved as if He were a man eternally condemned to Hell."[48]

Since, for Luther, God is both vengeful toward human sin and beyond any juridical scheme, the crisis becomes more aggravated. The curse—his wrath against the whole world—also conflicts with God's eternal grace in Christ, yet, because the blessing in Christ cannot be conquered, the curse must yield. What emerges is a disquieting image of God in conflict within himself.

This raises the question of how the Christian imagination has handled such a conflicted conception of God, whereby wrath must ultimately be taken for God's mercy. Two points are relevant. On the one hand, if God bore the conflict with sin within himself, believers could experience a more definite sense of salvation than in Anselm's scheme. On the other hand, this inner conflict found expression in church life through a relentless emphasis on the punishment Christ endured on behalf of individual sinners. The effect of this position is that it maintains God's wrath against a sinful world in full force while hovering latently and eternally behind God's love.[49]

Such a conception of God led inevitably to a chronic endorsement of divine violence as a core fixture in the Christian imagination at the

---

48. Luther, *Commentary on Romans*, 218, quoted in Bartlett, *Cross Purposes*, 90.
49. Bartlett, *Cross Purposes*, 93.

metaphysical level, disregarding the outpouring of limitless compassion by the Crucified in the abyss of suffering *at the hands of sinful men*. This attribute of Jesus' death should have given us pause long ago, integral as it is to the revelation of God as *Abba*, although such a conception differs considerably from what most Christians have been taught about God. This development in Western Christianity seems to have shut the door on the hope for wholeness and the experience of a completely unquestioning love, were it not for a new angle of view a Girardian reading offers.

The Revelation of the Forgiving Victim

So far, this chapter has attempted to engage Girard's mimetic theory as an instrument of revelation to unveil for us the inner constitution of the human person, the over-and-against character of individual and corporate identities, and thus the unavoidability of culture. We also saw how deeply all human knowledge and culture is marked by the consciousness of death. In the remainder of the chapter, my goal is to connect *this* human reality first with Christ's resurrection, then with the shape of his death and thus with the meaning of Christian salvation and its outworking amid the present human situation. This lays the foundation for what will be presented in the final part of the book: the need for an urgent recovery of God as *Abba*, who revealed himself in Christ as abyssal compassion and into whose all-encompassing subjectivity we are called from the discontinuity of our incidental individuality. This is not a conception of a God who all-powerfully rules over and providentially controls everything in his creation (as such a power would neither be creative nor compatible with freedom and love), but of a God who overcomes human violence persuasively in the weakness of a crucified man whom he raises from the dead.

To set the scene, let me connect the primal history of the Judeo-Christian tradition with Girard's thought.[50] In Girard's view, Israel's faith in *YHWH* evolved from an ancient sacral tradition in which orally transmitted myths were "gradually transformed in the light of new experiences of God."[51] Girard does not attempt to get behind the biblical text. His sole

---

50. Girard's theory is fully situated in the evolutionary paradigm and, due to its extraordinary explanatory reach, offers in my view a felicitous platform for articulating Christian doctrines for the modern age.

51. Schwager, *Banished from Eden*, 16.

aim is to disclose as fully as possible its mythical remnants. Strikingly, the Genesis story of primal history encompasses many elements of Girard's theory, as highlighted in this paraphrased summary:

- The story begins (in the garden) with a counterfeit imitation of God that leads to rivalry.
- The transgression of Adam and Eve takes place in the context of a prohibition.[52]
- The figure of evil appears (no longer as a god as in archaic religion), only as a creaturely agency (the serpent) with its role limited to that of the tempter.
- Human responsibility for sin begins with freedom of choice, yet human failure to act responsibly shows up in history as increasing violence, reaching its first climax when Cain murders his brother. With Lamech's 77-fold vengeance, violence escalates exponentially, and by the time the story gets to Noah violence has gone global.
- Gender relations and sexuality are all drawn into the vortex of mimetic rivalry (Gen 3:7, 16; 4:23f.; 6:1–3; 9:21f.).
- The tendency to offload personal culpability (evil) onto another (as in the scapegoat mechanism) is present from the start (Adam, Eve, Cain).
- Religious and cultural institutions like sacrifice and the city are mentioned, both in connection with violence.
- Mythological elements are present as well (a talking snake, giants, and undefined "sons of God").

Importantly, in biblical understanding God has nothing to do with human sin but gradually subverts and demythologizes the ancient sacred order, although it is not fully achieved in the Old Testament. *Torah*, the law, acting as instrument of discovery, detects personal and collective failure and sin so that Israel's foundational text, in contrast to ancient myths, shows the human condition for what it is. The violence and evil of the perpetrators are no longer veiled under the mystique of sacred violence. Instead, and contrary to the structure of ancient myths, God calls Adam and Eve to account; Cain's jealousy and his refusal to deal with it

52. In Girard's theory, the category of "sacred prohibitions" belongs to a later cultural/religious development of the scapegoat mechanism aimed at the prevention of the crisis (do not touch, take, eat, etc.).

are revealed (although God does not abandon the perpetrator despite his guilt).[53] Abel's death was neither fate (as in the death of Remus in ancient Roman mythology) nor necessary (as under the sacral order). Instead, Cain is identified as personally accountable.

As the Old Testament began the long process of demythologizing the ancient system of sacral violence, a new phase in human self-understanding opened that differed radically from that of the cultures surrounding Israel in the ancient Near East. More specifically, in Israel's imaginative remembering, its exodus from Egypt and the giving of the Torah ushered in a new social project that was not to be based on oppressive (pharaonic) methods of victimization. For, according to their faith, the Torah had been given as an expression of divine love along with the promise of a new future that inspired hope for the long term, not just for Israel but for the Gentiles as well.

However, as this people set out on their historical and God-accompanied journey, they had to discover repeatedly—under the demands of the Torah—how deeply flawed their own constitution was. To be sure, they had physically left behind Egypt's oppression, but not their own deeper enslavement to intramural rivalry and thus the consciousness of death. The latter not only harassed them in their brotherly relations but also seemed to be inescapable in other respects, as this death consciousness constituted the religious and cultural world around them.

If over the centuries prophetic expectations had nurtured a longing in the Jewish people for peace and stability from above in the fulfillment of God's promises, it was to remain hidden for several hundred years until the time of fulfillment had come, which in biblical understanding is associated with the coming of the promised Messiah. However, when this moment arrived in Jesus, the new turned out to be so radically unforeseen and different that it clashed with Israel's traditional understanding of God. After all, what God did in and through the crucified and risen Jesus was so unprecedented in Jewish conceptions that it eventually forced a substantial recasting of the traditional understanding among Jesus' first followers, who were all Jews.

In a masterful three-step exposition based on Girard's anthropology, James Alison—whom I follow here—undertakes a similar task for us, recasting our understanding of the God who raised Jesus from the

---

53. Alison perceptively notes that the story still retains an element of archaic arbitrariness in the way God accepts and rejects the sacrifices of the brothers (*Joy of Being Wrong*, 134).

dead.⁵⁴ In his first step, Alison engages Jesus' discussion with the Sadducees in Mark 12:18–27, emphasizing that even before his death Jesus was adamant that any understanding of God structured by death cannot speak adequately of God. After he was raised, it was possible for his disciples to see that God was indifferent to the reality of death in that God's love for the man Jesus was unaffected by his death. After all, it had caused no separation from God, as love was "being reciprocal even through death." For in God's sight "Death is as if it were not."⁵⁵ If God is the God of the living and not of the dead, then God has nothing to do with death, and anyone who thinks otherwise is "greatly mistaken" (Mark 12:27). Alison's conclusion is worth noting:

> The resurrection of Jesus, at the same time as it showed the unimaginable strength of divine love for a particular human being and therefore revealed the loving proximity of God, also marked a final and definitive sundering of God from any human representational capacity. Whereas before it could be understood that God did not die, nor change, nor have an end, this was always within the dialectical understanding of what does happen to humans. With the resurrection of Jesus from the dead there is suddenly no dialectical understanding of God available, because God has chosen his own terms on which to make himself known outside the possibility of human knowledge marked by death.⁵⁶

In other words, we learn from the resurrection how completely unaffected God is by death and how impossible it is for us to perceive God in his absolute freedom to love and make a dead man come to life in the midst of human history, universally structured as it is by the knowledge of death and its inevitability.

In the second step, Alison expounds the "content" of the resurrection: the man who was dead and now is alive. Since God loved this man Jesus, who was killed by other humans in a specific manner and from specific motives, his resurrection "confirmed his teaching and revealed the iniquity of what put him to death."⁵⁷ To bring this *factum* about, God involved himself in "one of us," demonstrating God's "extraordinary proximity to the human story," as Alison calls it, revealing simultaneously the distance between God and any human possibility of devising

54. Alison, *Joy of Being Wrong*, 115–19.
55. Alison, *Joy of Being Wrong*, 116.
56. Alison, *Joy of Being Wrong*, 116.
57. Alison, *Joy of Being Wrong*, 117.

adequate portrayals of God. In another way, the immanence of God in the life, death, and resurrection of Jesus also revealed God's complete transcendence in our history.[58]

By appealing to this paradox, Alison then develops a compelling logic for recasting our understanding of sin. The first step separated God from death and revealed the *human* nature of death—"whatever death is, God has nothing to do with it." On the other hand, through the resurrection we are made aware that death, although it is, need not be. Yet, it is also something more: while God has nothing to do with it, death is opposed to God, for if God wishes to make himself known in the resurrection of Jesus, then any representation of God that is other (say involved with death) is contrary to God. Therefore, as Alison writes, the death of the man Jesus, as death, is more than a merely biological reality; it is also a sinful one because we killed him:

> God, in raising Jesus, was not merely showing that death has no power over him, but also revealing that the putting to death of Jesus showed humans as actively involved in death. In human reality, death and sin are intertwined: the necessity of human death is itself a necessity born of sin. In us, death is not merely a passive reality but an active one, not something we merely receive but one we deal out.[59]

In the third step, Alison focuses on the fact that this man Jesus was also killed *for us*. Volumes have been written about the meaning of these two words, frequently to underscore the default position of Western Christianity that Jesus paid the price for our sins (see the previous section). As we continue with Alison's (Girard's) anthropological approach, we note that the resurrection also permits a recasting of God as nonviolent: God did not just raise Jesus because he loved him, but he raised the victim of human iniquity and violence as forgiveness, thus revealing God as love for us (John 3:16; Rom 3:21–26; 1 John 4:9–10).[60] If, then, the crucified and risen victim reveals God as love, it follows that the human reality of death, hijacked by sin, is capable of being forgiven. And further, if the raising of Jesus was to make forgiveness possible, then forgiveness must also extend into the very reality of human death, undoing our mimetically acquired knowledge of its reality and inevitability. This

---

58. Alison, *Joy of Being Wrong*, 117.
59. Alison, *Joy of Being Wrong*, 117.
60. Alison, *Joy of Being Wrong*, 118.

applies not only to our acts but to "what we have hitherto imagined to be our very natures." Alison concludes that if death is something that can be forgiven us, then "we were not only wrong about God, but we were fundamentally wrong about ourselves."[61]

Since none of these insights are available to us while our understanding of God and of ourselves is marked by death, they must be revealed to us through God's gratuitous love. Once we see this, we begin to realize that our consciousness need not be inflected by death. Since this is its condition, however, it is also true to say this matter is also the crux for an understanding of original sin. After all, our condition, constituted by universal mimesis and with a consciousness marked by death and thus contrary to God, is itself something capable of being forgiven. As humans-marked-by-death, the only way for us to appreciate our true condition lies precisely in what is revealed to us—namely, that this condition is unnecessary.[62]

In sum, then, because universal mimesis has blighted our very being, including our consciousness, with the knowledge of death through the scapegoat mechanism, all human efforts to make life secure through religion and culture are bent toward death and abide in death until undone from within through the revelation of the forgiving victim. In this revolutionary moment, human consciousness is reborn "to a living hope through the resurrection of Jesus" (1 Pet 1:3).

Because of our condition, it is a fundamental principle of human society that it misrecognizes even the possibility of an alternative, for this society crucified the one man who proclaimed it. Incapable of recognizing this possibility, let alone grasping it existentially, we need to be given "Easter eyes" to see it, which presupposes a thorough restructuring of consciousness from death to life, which only an encounter with God's limitless love can generate. That very moment of God's radical immanence also reveals God's utter transcendence, in whose presence death is no more. With this revelation and the return of the Forgiving Victim, the cosmos enters an entirely new phase in its history, as the humanized stuff of the cosmos receives for the first time an inherently nonrivalistic model to imitate, who, from the abyss of suffering and death, responds with abyssal compassion and new life.

---

61. Alison, *Joy of Being Wrong*, 118.
62. Alison, *Joy of Being Wrong*, 119.

Conclusion

The task of this chapter was to illuminate the paradox we humans are to ourselves. On the one hand, we are capable of abyssal interhuman dysfunction—myriad forms of conflict, brutality, and war. On the other, our prehistoric ancestors cared for the weak and disabled, attesting to an inherent capacity for authentic compassion and sacrificial love. The origin of this paradox, we discovered with René Girard, has to do with our constitution as imitative creatures, an attribute that sets our species apart from all others as it predisposes us to look to others for signals as to what is desirable, worth having, and worth striving for. Because this trait of *mimesis* is naturally acquisitive and conflict prone (two subjects may desire the same object), human social life is inherently fragile. What ensured the survival of archaic society was the evolution of culture and religion. The built-in mechanism of victimage (the scapegoat) lessened this fragility, keeping self-destructive outbreaks of unstoppable mimetic violence and vengeance in check and laying the foundation for the notion of sacrifice as a violence-channeling primordial cultural structure.

Still, throughout history it has been very difficult for humans to live in peace and harmony with each other. After all, as Girard explains, the signals and suggestions we have received from infancy have all been heavily weighted toward mimetic rivalry, so that from the beginning the identities of individuals, groups, and institutions have been shot through with conflict-prone mimesis, victimage, and expulsion. As an inward dynamic, this process generates phantom selves that must constantly distinguish themselves from others in an over-and-against fashion, often ending in bloodshed and brutality. Even the Christian church in its struggles for identity (orthodoxy) has not been exempt.

From here, by way of logical extension, the chapter turned to the God language that developed in Western Christendom from the fourth century onward. To demonstrate, five atonement theologies from Ambrose to the Protestant Reformation were compared with the divine image Jesus projected when speaking of Israel's God as *Abba*. Fascinatingly, we found that even the core of the Christian imagination—the theological interpretation of Jesus' death—did not escape an infiltration with mimetic rivalry, as Anthony Bartlett has shown. That all five theologies resort to typically mimetic categories in their interpretations speaks of the depth to which the human constitution has shaped even our understanding of God, projecting human violence on the transcendent screen in a chronic endorsement of divine violence. But preserving aspects of the primitive scared came at the expense of the outpouring of limitless compassion from the abyss of Jesus' suffering, so that even in the Christian era all human knowledge and culture remained deeply marked by the consciousness of death.

The last section of the chapter sought to connect the resurrection of Jesus with this reality. Starting with insights from Israel's primal history, we noted that the Torah no longer viewed violence and evil through the mystique of sacred violence (as happens in ancient myths) but revealed what they really are in the human situation. In other words, the Old Testament began to demythologize the ancient sacred order, leading to the conclusion that *YHWH* has nothing to do with human sin but instead subverts it from within. Although not fully achieved in the Old Testament, this trajectory was nevertheless presented as the expression of divine love and the promise of a new future, not only for Israel but also for the Gentiles. In the closing section, we briefly explored an unfamiliar notion: when God did fulfill his promise through the life, death, and resurrection of Jesus Christ, this new act was so different from Israel's tradition-bound expectations that they could not receive it.

This urges us to pay close attention to this historical reality when considering what the resurrection signifies theologically: that from now on (after Jesus was raised) any representation of God that involves death is inadequate. After all, Jesus' death did not result in separation from God, a reality that invites us to dare a recasting of God as utterly nonviolent. This flies in the face of all human understanding of God and of ourselves as these remain marked by death. Yet, if the resurrection is true, this inflection is unnecessary. Universal mimesis has blighted our very being (accounting for the paradox that opened the chapter), a condition that includes every effort to render life secure through culture and religion. But the testimony of the New Testament proclaims that any inflection toward death can be undone from within through an encounter and ongoing companionship with the Risen and Forgiving Victim.

∞ 9 ∞

Toward an Expanded Theology

I CANNOT THINK OF better words for the beginning of the last chapter than those attributed to Sir Isaac Newton: "I see myself as a child playing on the seashore, and from time to time he enjoys discovering a more polished pebble or a more beautiful shell than usual, while before me the immense ocean of truth spreads unexplored."[1] It would certainly be absurd to consider Newton's colossal contribution to mathematics and physics as "child's play." However, when he compared his achievements with the vast ocean of unexplored truths, he could not help but regard his work with humility and modesty.

Of Seashells and Pebbles

Perhaps you too have, in the preceding chapters, picked up a pebble or a seashell here and there that spoke to you more significantly than others about the vastness from which they have come, or about the indescribable interrelatedness and beauty of God's creation, of the Creator's grandeur, and the extraordinary depth to which God in Christ has gone to reveal his true character. Perhaps you too have been reduced to silent adoration in the presence of God's wisdom. Perhaps it has inspired you to walk more often with open eyes on the beaches of God's creation,

---

1. Cited by Funes et al., "Searching for Spiritual Signatures."

possibly realizing for the first time that you yourself are an integral part of it. Perhaps it has dawned on you that you are much more deeply part of the creation than you have ever before realized, and that despite its remarkable intelligibility this creation is incomplete, still awaiting fuller realization in the future. If that is the case, my effort in guiding this beach walk will have been amply rewarded.

Yet, this walk itself is still incomplete. We must still ponder some of the pebbles and seashells we have gathered for their meaning in the attempt to cast a theological vision expansive enough to encompass what has been brought to light by the scientific description of the cosmos. In our approach to this task, I propose to reflect theologically on several subtexts from the preceding chapters, such as the shift in perception of the natural world from fragments to wholeness, the emergence of subjectivity in a purportedly mindless universe, God as Trinity in creation and evolution, the rediscovery of God as *Abba*, and Jesus as the new structure of the human. In short, what this chapter offers is a set of theologically framed windows looking out over the landscape of the evolutionary movement we call cosmic history in light of the revelation of God as love crucified and risen.

As we consider these themes, it is quite possible that new issues may insert themselves into our consciousness—for instance, how this new world picture might affect the way Christianity has traditionally defended its claims. I mention this point because, in an age of science, the Christian tradition cannot escape some critical criteria if it wants to be heard in the public square. These derive not primarily from discoveries in the disciplines themselves—in physics, cosmology, biology, genetics, and so on—but from the inner logic of critical inquiry and the nature of public discourse in our time. Let me mention five:

1. Those offering public truth claims are required to justify their validity.
2. If truth claims are to be public, they must be universally accessible and open to discussion.
3. If truth claims are open to discussion, the notion of absolute validity must be abandoned, for discussion can only lead to relative truth that will always reflect the historical situation of its time.

4. If truth claims are to stand up to historical-critical inquiry, their research base must be universally accessible; it cannot belong to a privileged in-group with authority over its communication.

5. Because public truth claims must remain communicable, no tradition can convincingly claim that only its adherents can understand what that tradition is about.[2]

If Christianity is to regain a measure of cultural influence, it will need to show how its core symbolism and message are up to this challenge in an idiom that resonates with its cultural setting. The summarizing reflections and observations of this chapter are offered to further this end. On the flipside, strategies of communication that claim for Christianity divinely favored-nation status or immunity from critical inquiry may turn out to be self-defeating.

Wholeness

With the Copernican revolution and the advance of the natural sciences—as we saw in the preceding chapters—a new picture of the world emerged. Earth and humanity were no longer central, while science made it progressively possible to explain much of what was previously attributed to God's role in the universe. When this new world picture merged with Descartes's dualism, with its focus on the self-thinking subject, it impressed on Western culture, despite its Christian roots, a sense of profound separation from God and nature. Intellectually, this milieu became fertile soil for Newton's proposal of a divinely created clockwork universe that ran perfectly on the principles of mathematics and mechanics. The contrast to prescientific perceptions could not have been starker. Before the rise of science, all natural reality was regarded as being alive, from rocks to rivers, from plants to animals, even sky and sun. Now Newton's distant God, the great engineer, had created an inert and machine-like cosmos that was hostile to life and ill-suited as a medium for personal communication between God and humans. This conception of the world mediated not life but alienation and indifference. Yet, from

---

2. Admittedly, outsiders to the faith will be unable to enter the same experience of living the faith and benefit from its promises.

the Enlightenment onward, such was the perception of the natural world, the reigning paradigm for the next 300 years.[3]

To be sure, in Newton's day the Creator was still believed to be behind the laws of physics, but this deity was far away, outside of time, while the expanse of space was perceived as no more than a cosmic container randomly filled with inert stellar objects. Unsurprisingly, this pessimistic cosmological paradigm imbued the consciousness of Western culture with a deep sense of separateness and fragmentation that is still with us.

Yet, by the nineteenth century, seismic changes were on their way that would shake this mental fortress to the core, awakening the possibility of a more unified view of the material world. James Maxwell was the first to discover a field theory when he unified electricity and magnetism as electromagnetism; Charles Darwin published his now widely accepted theory of speciation over time, a kind of field theory in biology; Max Planck revolutionized atomic and subatomic physics with the discovery of the quantum world; and Albert Einstein struck a death blow to Newton's clockwork universe by bringing mass and energy into a new dynamic equivalence through the most famous formula of all time, $E = mc^2$. When, in the early twentieth century, Einstein redefined gravity as the curvature of space-time, he opened the door to a new understanding of the cosmos as an internally dynamic whole. Edwin Hubble's discovery a few years later of galaxies besides the Milky Way delivered earthshaking observational proof that, rather than being a static container, the universe was an expanding and moving cosmos.

Working at the other end of cosmic size spectrum, a group of great theoretical physicists—Niels Bohr, Max Born, Werner Heisenberg, Pascual Jordan—were tackling the interpretation of the quantum theory at the same time. Their conclusion was equally consequential: the realm of the quantum, the deepest foundational layer of material reality, could no longer be understood independently of the mind of the observer. This verdict negated once and for all Descartes's dualism of mind and matter, opening the door to the view that subjectivity can no longer be excluded from the foundation of existence. Now, wherever we looked in

---

3. Contemporary scientific and philosophical thought is profoundly attracted to a materialist view of the world. However, its devotees can assign no meaning to what exists because, in their conception, the material world is constituted by dead and mindless matter. This widely held belief blinds them to an inherent incoherence of this argument: How can dead and mindless matter enable them to produce reliable conclusions?

the universe, from the largest cosmic structures to the most complex biological systems down to the quantum realm, the natural order proved to be not only intrinsically on the move and deeply interconnected but also hospitable to subjectivity. By the mid-twentieth century it was no longer a secret that the old paradigm of separation and fragmentation was on the way out.

Since the cosmic process itself seemed fine-tuned for life, implying wholeness at its core, so-called objects of the natural world could now be better understood as the external manifestations of a far more intimate interplay occurring at levels where wholeness is primary. In that case, instead of being dialectically related as dualism assumes, matter and mind—since they have appeared in human beings—may reflect a foundational reality of the natural order. Sir James Jeans, working in the field of quantum physics, had suggested the idea in the 1930s that conceiving the universe as mindlike may not be incompatible with a quantum universe.[4] In the 1990s, Stuart Hameroff—together with the 2020 Nobel laureate in physics, Roger Penrose—developed this line of thought further, suggesting that a mental nexus exists between mind and matter through which evolutionary novelty could occur. Starting with the question "What is empty space?" (a question already debated in ancient Greece), they showed that, according to Einstein's theory of general relativity, space is "a richly-endowed plenum," the space-time metric. At very small scales, however, space-time is not smooth but "quantized" or granular, involving finely woven networks of quantum spins. Penrose believes that these small and weak processes could be linked to biology.[5]

Theoretical physicist David Bohm conceptualized this inherently dynamic and interrelational view of material reality in a more determinist view of the quantum world. He suggested that throughout the universe interactions occur through "pilot waves" that guide interrelational outcomes at the quantum level; these pilot waves were pointing to what Bohm understood as an "implicate order" (see chapter 4). On this view, material reality could no longer be explained from external interactions between things but only from internal or "enfolded" relationships governed by information content. In other words, mind and matter are of

---

4. Jeans, *Mysterious Universe*, 158.

5. This occurs at the Planck scale of 10^{-33} centimeters and 10^{-43} seconds (Hameroff, "Funda-Mentality," https://quantumconsciousness.org/content/funda-mentality). See also Penrose and Hameroff, "What 'Gaps?'"; Hameroff and Penrose, "Reduction of Quantum Coherence."

the same order, and the primary reality belongs to the whole rather than to the parts. Recognizing that the relationship between matter and mind (consciousness) was blatantly lacking in modern science, Bohm took that relationship seriously in his pursuit of the whole. For him, the whole was not bounded by space-time; it was the creative process of its totality that could not be expressed in the four dimensions of space and time. Thus, the whole was beyond appearances and beyond scientific experiments.

On reflection, in such a universe the arrival of human beings can no longer be understood as an evolutionary fluke but, like everything else, as integral to the whole. Scientists who think this way reject the materialist claim that the universe is lifeless and mindless, not least because of the extraordinary complexity and neural connectivity observed in the human brain. For here the stuff of the universe has appeared (as far as we know) in its most synthesized form, attaining the capacity for conscious self-reflection. Hence John Haught's proposal that the epic journey of the universe is, apart from atoms, molecules, cells, and social groups, also about subjectivity.[6] As I see it, it is time for Christianity to come to terms with these realities of the natural world and for creative reflections on theological implications.

The next section reflects on the emergence of subjectivity within cosmic history. Taking the cue from the observation that the universe includes some form of subjective experiences in all living creatures, one must admit the extraordinary possibility in an evolving universe that mind or consciousness must have been present potentially in the big bang itself.

Subjectivity

For centuries, it has been a matter of intellectual correctness in the natural sciences to affirm that all matter and therefore the universe itself is mindless, a notion that has enjoyed almost universal acceptance since the days of Descartes. Descartes's view had divided the world into two substances, *res extensa* and *res cogitans*—inert matter and the soul. *Res extensa* belonged to the realm of corporeal things, could be described by logic, and was characterized by definiteness. *Res cogitans*, by contrast, referred to the less tangible realm of mental events like thoughts, desires, and feelings. This separation solidified into an axiom. *Res extensa* emerged as the

---

6. Haught, *New Cosmic Story*, 3.

exclusive focus of the natural sciences, banishing, as a matter of principle, concerns with subjectivity or inner experience from the purview of legitimate scientific inquiry. The rationale was simple: only dead matter can be measured, weighed, and otherwise manipulated. But if matter is "inert," then all processes in the universe, including the processes of life, are reduced to mere material, even mechanical events. More importantly, so the materialist argument goes, an inert cosmos makes nonsense of attributing meaning and purpose to the universe, let alone resonance with divine revelation, if it exists. Therefore, a theological vision can only make sense if it can be established that subjectivity and inner experiences are indeed integral to the cosmos.

As we saw in chapter 3, planet Earth, together with billions of other minor cosmic bodies, is an aggregation of materials that were once produced inside stars. Yet, the remarkable feature of this planet is that, as far as we know, it is the only place where subjectivity and consciousness have evolved. Philosophers and other scholars concerned with this phenomenon have long debated the "hard problem of consciousness." Coined by the Australian philosopher David Chalmers, this term refers to the difficulty we encounter when we try to explain how an entity entirely made of physical matter, such as a human being, can experience, for instance, the redness of a rose or the wetness of water, and even be conscious of its own existence. This issue has also become a crucial element in the science-religion dialogue. After all, for example, the argument that mind is fully natural or a fundamental property of the material world does not, at first glance, sit easily with either science or Christian theology.

Yet, as Joanna Leidenhag has shown, this hypothesis is undergoing a significant revival among philosophers of mind as well as among theologians.[7] While space does not allow us to examine it, I would like to suggest, in keeping with the perspective of this book, that I regard the inward dimension of the universe as an integral aspect of evolutionary cosmic advance, for two reasons. First, the role of God in the natural world cannot be meaningfully discussed as long as the sciences reject subjective experience as part of the explanation of what happens in evolution; hence, legitimating the presence of subjects is an essential precondition for any religious/theological view. Second, with the arrival of *Homo sapiens*, self-reflective subjects have appeared convincingly in the evolutionary cosmic process.

7. Leidenhag, "Revival of Panpsychism." Anyone interested in this subject will find much food for thought in Leidenhag's article and in the literature she cites.

In the remainder of this section and in the next, I will focus on the gradual awakening of subjectivity in the cosmos, expecting to offer further insights when we look at the process of creation from a trinitarian perspective.

Before the cosmic process could achieve subjectivity in its fullest human sense, life first had to arise somewhere on the planet.[8] Life as we know it, even at the lowest level of organization, is animated by a mysterious energy that relentlessly strives for self-expression with the tacit anticipation of reaching higher levels, as the philosopher Hans Jonas has argued. In other words, with the appearance of even the most primitive forms of metabolism, an "inner sense" has entered cosmic history.[9] Jonas argues that even these early forms of life "embryonically" contain, at a cellular level, all of the experiences of higher forms of organic organization, including such feelings as "being torn between freedom and necessity, self-sufficiency and dependence, connectedness and isolation." Such themes, he maintains, simply belong to all life and are traceable "in an ascending order of organic capabilities and functions," reaching their pinnacle in the human being.[10] Convinced that "inner experience" was always a cosmic possibility, Jonas holds that it belongs to the actuality of life to reach upward from the "most primitive stirrings of life" to the human level of organization.[11]

By positing the presence of subjectivity at the lowest possible level of organic organization, this view counteracts the mechanistic interpretations of life. In philosophy this issue was not a new problem. Hungarian-born British polymath Michael Polanyi had addressed it forty years earlier in his work *Personal Knowledge*. Polanyi, affirming subjectivity, argued that all knowledge—even tacit knowledge—was personal knowledge, enabling us to instinctively recognize traits identical with our own like "striving, achieving, and failing" in other living beings.[12]

On the other hand, in their commitment to impersonal categories, the natural sciences are unable to acknowledge these traits as emanating

---

8. Earth may not be the only planet where life has emerged in the universe. According to a NASA study, there are 300 million potentially habitable planets in the Milky Way galaxy alone (https://newatlas.com/space/300-million-potentially-habitable-planets-milky-way-galaxy/).

9. I owe to John Haught my awareness of Hans Jonas's work *Mortality and Morality*.

10. Jonas, *Mortality and Morality*, 60.

11. Haught, *God after Darwin*, 169.

12. Polanyi, *Personal Knowledge*, quoted in Haught, *God after Darwin*, 168.

from a subjective core, despite the fact that modern neuroscience regards the human brain as a subjective (personal) meaning-making organ that at every moment turns millions of sensory impressions into a coherent picture of the world around us (see chapter 6). In other words, accepting subjectivity as an objective fact of nature, at least since the beginning of primitive life forms, is by no means unreasonable.

But what of the presence of subjectivity in ordinary matter—that is, in matter prior to the emergence of biotic life? On reflection, behind this question lurks another: Can it be reasonably claimed that all matter was hostile to subjectivity during the prebiotic stage of cosmic history? If the answer is yes, we are plunged once more into the void of mindlessness despite the rich evidence for subjectivity in the cosmos today. The depth of subjectivity we experience today suggests that at no time during the epic journey of the universe has there been a stage when the connection of matter with mind reached absolute zero. Instead, Jonas argues, subjectivity has always been present even though in a state of "latency" or "mind sleep," which is a far cry from the wholesale mindlessness asserted by scientific naturalism.[13] But latency simply means that the preanimate phase of the material world included the potential for the emergence of subjective experience in metabolic forms of existence. On this reasoning, when metabolism appeared, a clear line of demarcation between the two realms of "mind sleep" and "mind wake" was drawn in the sands of time. It identified the unique and unimaginably momentous historical occasion when the entire universe crossed the Rubicon, leaving behind mind sleep and stepping into a state of *cosmic awakening* to subjective experiences.

As plausible as this answer may sound in the ears of the philosopher, the theologian must ask more specifically, "Where is God in evolution?" and "How can God be effective in cosmic history ongoingly when vast regions remain in their nonmetabolic state?" But before we consider a Christian approach to these questions, let me briefly show the antisubjectivity bias among natural scientists while clarifying a semantic ambiguity.

So far, I have used the terms *subjectivity*, *inner sense*, and *consciousness* synonymously, downplaying the distinction, for instance, between subjective experiences of animals and humans. Strictly speaking, human consciousness is always to be understood as self-reflective consciousness,

---

13. Haught, *God after Darwin*, 170.

meaning that humans are aware of being conscious and that they can reflect on this state of mind, something animals do not do.

By the same token, there can be no doubt that animals do possess a core emotional system that is able to generate feelings of care, fear, play, panic, rage, even pleasure seeking, all of which all animals share, and many animals display these in symbolic behavior such as gestures and dances, as we noted in chapter 7. Sadly, Western culture has been exceedingly slow in taking this reality of the natural world seriously. Aristotle held that animals—to secure their survival—were endowed with instincts, but only humans possessed rational souls. Two thousand years later, Descartes, true to his mechanistic philosophy, treated animals as "reflex-driven machines."[14]

It took another 200 years, with help from Darwin's theory, for scientists to become more inclined to grant some continuity between humans and animals. Thus, toward the end of the nineteenth century, psychologist William James (1842–1910) became convinced that consciousness occurred with differing intensities across the animal kingdom. However, it was only in the last few decades that the natural sciences began to lay the groundwork for examining questions of animal cognition and consciousness seriously. This in no way means that studies of animal consciousness now enjoy wide-ranging approval among scientists; the topic remains controversial, even taboo.[15] What is critical for our purposes is not so much how consciousness works in human and nonhuman animals, but that consciousness has evolved in cosmic history on a broad scale, pointing to rising cosmic inwardness.

Inwardness

In earlier chapters we connected this rising inwardness with the potential for an openness to divine revelation, positing a feature of the creation that bears in itself the potential for expressing in the finite domain what belongs to the infinite. Inwardness may therefore be regarded as a relational attribute enabling the potential bonding of the Creator with the creation—the infinite with the finite, the absolute with the relative. If this were granted, inwardness is by design imbued with a tilt toward the

14. "Animal Consciousness," https://stanford.library.sydney.edu.au/entries/consciousness-animal/index.html.

15. "Animal Consciousness."

Creator, for it cannot be qualitatively and structurally inconsistent with his ways of acting, namely, in freedom and love. The writer of Ecclesiastes puts it this way: "[God] has made everything beautiful in its time; also he has put eternity in to man's mind" (Eccl 3:11). Hence, inwardness so understood cannot belong to the order of cause and effect, for the infinite cannot be a member of the causal order; nor can it lend itself to coercive or controlling ways without corrupting by curtailing freedom; it cannot even be explanatory, for the finite order is limited to finite descriptions. What I am implying here is that inwardness must be conducive to what is self-expressive, self-authenticating, self-sharing—that is, to an order that is incarnational.

By embracing the incarnational metaphor in the finite realm, we cannot help but first point to the inner order nature displays. Let me illustrate it with examples from recognizable patterns of mathematical relations in nature like the Fibonacci sequences,[16] the golden ratio,[17] Euler's law,[18] or the Fourier transform.[19] All of these point to a structural affinity between what the human mind is capable of conceiving and its tangible counterparts in the natural world. The incarnational metaphor also calls us to come to terms with phenomena that occur in the natural world as manifestations of the world's constitution, such as the laws of thermodynamics and mutation-producing cosmic radiation; gravitational synchrony capable of dislodging large objects from the asteroid belt and hurling them toward Earth; tectonic movements in the Earth's crust

---

16. The Fibonacci sequence is a sequence of numbers where each number is the sum of the two preceding ones, starting from 0 and 1. They appear in many natural configurations—in the branching of trees, the way leaves appear on stems, the arrangements of fruit sprouts of pineapples, the flowering of artichokes, and the way pine-cone bracts are arranged.

17. The golden ratio, also called the golden mean, golden section, or golden proportion, has been studied by mathematicians ever since Euclid. It has been used to analyze the proportions of natural objects as well as human-made systems. It may be observed in some natural patterns like the spiral arrangement of leaves. Artists and architects use it in their designs to achieve aesthetically appealing proportions.

18. Euler's mathematical formula creates a powerful connection or fundamental relationship between two branches of mathematics—that is, analysis (calculus) and trigonometry.

19. A Fourier transform relates the frequency domain (electromagnetic frequencies of the light or sound spectrum) to the domain of space and time (what we see and hear). It is a mathematical operation that breaks down a function (of time) such as a signal into its constituent frequencies, or a musical chord into the volumes and frequencies of its individual notes.

that unleash earthquakes and tsunamis; population explosions, diseases, and epidemics; pain, predation, and death. Since all of these have existed as part of the inner structuring of the created world long before humans walked on the planet, Christians need to come to terms with the fact that attributing them to the moral failure of a primordial couple is no longer coherent.

Further, looking at the world from the standpoint of science, we find that we humans are not just part of the cosmos in a macroscopic sense but more profoundly than our imagination ordinarily suggests. Made of the same atoms as butterflies, alpine mountains, and stars, what constitutes us at the level below the atom is the same restless swarming of quantum happenings that occur at the foundation of the created order—the fleeting appearance and disappearance of transitory subatomic entities, a world not of things but of random events.[20] If, theologically speaking, the universe is derived from an outpouring of infinitely creative freedom-granting love, then the universe itself must be inwardly imbued with creative powers able to bring forth new nonliving and living wholes at all levels of cosmic organization (physical, chemical, biological, social). This in turn will feature creative propensities with the freedom to try all realizable possibilities, without prior planning and external control.

If this is so in a finite creation, energy inputs and raw materials that sustain these processes can only come from within the cosmic ecology itself. To bring forth higher orders of organization, our cosmos must rely ongoingly on the conversion of its own substance in all processes, from galactic birth and star formation to the food chain all the way down to cellular metabolism. In other words, in a finite, incomplete, and self-organizing creation based on a describable and implicate order, the disintegration and recycling of decay products are as unavoidable as feeding, ingestion, excretion, decomposition, predation, and dying. At the same time—with the emergence of human consciousness and from within the same creative processes—certain attitudes, values, and behaviors reflecting a growing awareness of "rightness" that also belongs to cosmic inwardness appeared in cosmic history very, very recently.[21]

As we saw in the three preceding chapters, human consciousness was not an empty container just waiting to be filled. Rather, for hundreds

20. Rovelli, *Seven Brief Lessons*, 64.

21. "Rightness" is a dimension that, according to Haught, humans "cannot grasp . . . directly but may have an awareness of being grasped by it" (Haught, *New Cosmic Story*, 114).

of thousands of years prehuman consciousness as it emerged in our hominid ancestors had resonated deeply with the natural world, its soils and seasons, its creatures and climates, including nature's awesome powers to bring forth life but also ravage it. To the degree that our ancestors experienced this transcendent dimension, or the sacred, they experienced it as the *mysterium tremendum et fascinans*, the dread of unpredictable and uncaring outbreaks of nature's destructive powers, but also as an expression of their abounding capacity for renewal and fecundity. We recall from chapter 7 how fascinated our ancestors were with these untamable forces of nature. Even much later, in the Minoan culture (Crete, about 4,000 years ago) and elsewhere in the ancient Near East, for instance, this fascination expressed itself in the highly evocative symbol for the powers of fertility as well as of destruction—the wild bull—just as it had expressed itself 20,000 earlier in the bear cult of the Cro-Magnon people.

Presumably, in a consciousness structured by the experiences of an ambivalent yet inexorable reality, inklings of rightness would have arisen slowly. Yet, as they did, our ancestors developed cultural artifacts of rightness like prohibitions, rituals, and sacrifices to avert ever-threatening possibilities of misfortune and to garner favor from these unpredictable powers, which were personalized at the same time. That such a deep existential link existed between the experience of awesome dread and the emergence of rightness is profoundly retained in Israel's memory—the awesome theophany at Mount Sinai that preceded the giving of the Torah (Exod 19:16–20):

> On the morning of the third day there was thunder and lightning, with a thick cloud over the mountain, and a very loud trumpet blast. Everyone in the camp trembled. Then Moses led the people out of the camp to meet with God, and they stood at the foot of the mountain. Mount Sinai was covered with smoke, because the Lord descended to it in fire. The smoke billowed up from it like smoke from a furnace, and the mountain trembled violently. As the sound of the trumpet grew louder and louder, Moses spoke and the voice of God answered him.

Feasibly, the more a sense of rightness and self-conscious reflection emerged, the opportunity for noninstinctive behaviors was born, which fostered community cohesion, self-sacrifice, compassion, even expressions of beauty. After all, such traits are already present in rudimentary form in the animal world, albeit instinctively. For instance, an Australian owl-like bird, the tawny frogmouth, is known for engaging in deep mourning over the loss of its mate. It will sit for days near the site of its mate's death, going without food, sounding like a baby.[22]

Those who are critical of evolution for its reliance on deadly competition, randomness, aloofness to suffering, and wastefulness are reminded that generative adaptive changes in the history of living creatures would have never occurred had the processes been only painful. Admittedly, these features do exist, and if they offend our own sense of rightness, then it is only because in us, as Haught has argued, the stuff of the universe has awakened to a deeper current of subjectivity.[23]

This inwardness, out of sympathy with suffering creatures and against the pressure of natural selection, counts excessive anguish and waste as somehow wrong. If this is so, the gradual awakening of capacities such as empathy, sympathy, and compassion must be considered as much a part of the cosmic journey as everything else, leading eventually to the appearance of *Homo sapiens*, where this awakening is still a work in progress. This is not to say that these values are nothing but products of nature. According to the Christian faith, they spring from the infinite and may be incarnated in the finite, perfectly fused eventually into a living unity in Jesus of Nazareth, the central revelation of God in Christianity.

Before elaborating further, it is opportune to visit briefly some of the philosophical commitments that tend to hold Christianity back from engaging constructively with the discoveries of modern science.

Christian Inertia

For almost 1,000 years, traditional theological reflection had conceived of God's influence on the creation in terms of Aristotle's notion of the "Prime Mover." This conception placed God in the role of a chauffeur who drives creation like one drives a car, from the past toward the future, acting providentially all the way. That this view is deeply embedded in the Christian tradition cannot be doubted. Yet, close attention to the picture provided by modern cosmology and biology urges us to think differently, as Teilhard de Chardin observed a century ago. In this new story it is feasible to understand what we call evolution as God's promissory bestowal of an externalizing dynamism open to its final consummation, a potency inherent in the material world that is behind the observable movement

22. Heathcote, "Tawny Frogmouths," para. 5.
23. Haught, *New Cosmic Story*, 169.

toward greater depth, relationship, union, life, and being. As we will see in reflections on the Trinity, God's all-pervasive presence is not just behind but also always up ahead, drawing the creation toward a future of ultimate union with God's "glory" (Ps 8), where one day Christ will be "all in all" (Col 3:11).

There can be no doubt that Christianity in general has been disinclined to engage with the discoveries of modern science. The reasons are both philosophical and ideological. There is, first, Plato's and Aristotle's metaphysics of *being*, which dominated theological discourse for centuries. These views led to the hierarchical picture of the cosmos that we find traditionally in Christianity, Judaism, and Islam. This being-centered metaphysics tends to obscure one of the most obvious aspects of nature, that nature is *becoming*. What's more, the notion of becoming presupposes an understanding of reality that emphasizes the importance of the future. If we understand God's love as the force that draws creation ever forward toward "a deeper coherence by an ultimate force of attraction" (which Teilhard de Chardin abstractly referred to as Omega), it would open up a vision of the divine dwelling not above but up ahead, as the ultimate goal of a world still in the making.[24]

Here much care is needed to avoid an inadvertent slip into Platonism, a stance that regards the present creation as a deviant and corrupted version of another world existing in primordial perfection above in the eternal present, untouched by time. Since this view is based on a metaphysics of the past, it tends to foster an unjustified longing for a lost paradise instead of the profound sense of expectancy needed by faith in the resurrection and in a new creation. Moreover, a theology infiltrated by a metaphysics of the past will resist the weighty sense of the future demanded by an evolutionary understanding of the cosmos that sees the entire creation as a work in progress, including us humans as cocreators. In other words, to the extent that Christian thought remains tied to such a metaphysics, it will distance itself from any meaningful engagement with the discoveries of science on questions of the origin and ultimate destiny of the cosmic process.

Unwittingly, this stance tends to play into the hands of the present cultural milieu of Western civilization, where atheism is the default position, and where the notion of a Creator and even the word *God* have become offensive. The usual critique first indicts Christianity's history,

24. See also Moltmann, *Theology of Hope*.

charging that those claiming to worship the Christian God have been responsible for much pain and suffering on account of religious violence and mass murder. Since these accusations cannot be denied, Christianity must admit that its historical record is not pretty, that its conduct has contradicted the model of Jesus on countless occasions over long periods and has even justified these deviations theologically. As discussed in chapter 8, from Ambrose to Anselm and Gregory the Great via Augustine and then all the way to the Reformation and the present age, the church has interpreted Jesus' death in terms of divine violence, even as it proclaimed in the same breath God's sovereign grace. In that chapter, I called this development (following Anthony Bartlett) "the crisis of God as *Abba*." In the next section, I will swim an extra mile against this current, as I attempt to recover an appreciation of the God I believe Jesus incarnated and proclaimed, the God of bottomless compassion, whom he called *Abba*.

Abba Incarnate

While it has always been true that Christianity is about God's relation to us and our relation to God, what individual believers and the church have understood that to mean does not always align easily with what Jesus actually modeled and, presumably, how Jesus experienced the Father's presence in his life. Given the many occasions in the Gospels where Jesus asserts his identity with the Father, it is wise to pay fresh attention.[25]

Jesus was brought up in Nazareth, a region of Galilee that had been captured by the Romans in 37 BCE. At that time, Galilee and Perea were ruled by Herod Antipas, the son of Herod the Great, who had died in 4 CE. Continuing his father's ambitious building projects, Antipas imposed heavy taxes on the local peasant population, forcing many into land forfeiture and abject poverty.

When Jesus stepped into history, he found a society split in two—a wealthy land-grabbing elite backed by Roman state power and ordinary people traumatized by both. No wonder multitudes thronged around him—hungry, distraught, and sick. The first three Gospels reveal his response: "He was moved with compassion" (Matt 9:36 and cognates). At the beginning of his ministry and identifying personally with the prophetic voice of Isaiah, Jesus proclaimed in Luke's Gospel (Luke 4:18–19):

---

25. For example, Matt 11:27; John 5:24–30; 6:38–40; 8:16–18; 10:15–18; 17:1–26.

"The Spirit of the Lord is upon me, because he has anointed me to preach good news to the poor; he has sent me to proclaim freedom for the prisoners and recovering of sight to the blind, to release the oppressed, to proclaim the year of the Lord's favor." He continued, "Today, this scripture is fulfilled in your hearing" (Luke 4:21).

It was to the little people primarily that he declared the nearness of "the kingdom of heaven." Those who would enter it were exhorted to act as if it had already arrived in fullness and they had been grasped by what was to come. In that kingdom the poor would be first and the rich last; the lowly would be exalted and the mighty brought low. It was a peaceable kingdom whose members had rejected the vicious cycle of reciprocal violence and replaced it with a response toward oppressors that empowered the oppressed: they prayed for their persecutors, they forgave an offense, when slapped they turned the other cheek, and when forced to walk one mile they walked two, all out of love for God and neighbor. The doorway to that kingdom was the present moment. Entry did not depend on holding to the right creed but on trusting Jesus's transforming message and acting on it. Then the hungry would be fed, strangers would be welcomed, prisoners would be visited, and even enemies would be loved. Yet, preaching and acting out such an agenda, with its implicit critique of imperial power under the gaze of Rome's legions, was both revolutionary and dangerous for Jesus and his disciples.

Thus, in conduct and teaching Jesus stood in the tradition of Israel's prophets. From Amos onward, they had frequently spoken like *YHWH*'s prosecuting attorneys against the disloyalty of his chosen people, especially their rulers. The prophets issued harsh warnings of the painful consequences that would follow if they did not live with integrity what they professed religiously. It was people's hypocrisy that most often riled the prophetic ire as they appeared to be religious but failed to deal with unjust social structures. Obedience to God was not to be seen in the opulence of temple sacrifices but in the depth of their compassion toward widows, orphans, and other underprivileged people. For the God of these prophets, whom Jesus presented as *Abba*, had made Israel's poor his special concern. What *Abba* was after was not outward religious observance, which can only produce blind, fear-centered self-concern, but a new way of seeing our neighbors as precious and infinitely loved members of the same humanity.

Jesus' understanding of God cannot be disconnected from his childhood experiences in his home at Nazareth, in the quiet household of

Joseph, the carpenter; nor can it be divorced from the words of his mother or from the Jewish piety and tradition that marked his early life, including the annual Passover pilgrimages to Jerusalem; nor can it be separated from the scriptures he studied. Yet, this grounding notwithstanding, one wonders whether it sufficed to impart to him the deep intimacy and tenderness that so centrally characterized his knowledge of *Abba*. From the Gospel texts we must assume that his knowledge of the God of Israel and his own identity were rooted in something supremely more personal, the uninterrupted and undiluted experience of *Abba*'s presence.

The one text that most explicitly speaks of Jesus' *Abba* consciousness comes to us from Mark 14:36, while the Gospel of John (5:19 – 20) affirms his utterly self-sharing participation in it: "I tell you the truth, the Son can do nothing by himself, he can do only what he sees the Father doing, because whatever the Father does, the Son also does. For the Father loves the Son and shows him all he does." Nothing less can explain that distinctive mark of his work, so openly on display in his interaction with the multitudes and later in his extremity—the compassion that moved him as he forgave their sins, healed the sick, fed the hungry, and showed them how to live fully and dedicatedly under the yoke of Roman oppression, even unto death.

In chapter 7, I used the image of a father bending over the crib of his newborn baby. While conveying something of the deep and tender sympathy a father feels for his child, this image is too limiting in two respects. Yes, God is personal, and so much more than any human father can ever be; indeed God is inexhaustibly personal. Yet, for all the personal warmth of this affirmation, we must guard against a common misconception: the Aramaic word *Abba* does not imply that Jesus uttered baby talk when he referred to God the Father.[26] Rather, the word draws us toward the all-committing personal and infinitely self-donating love that characterizes God's identification with his creation in Jesus and Jesus' identification with the absolute self-gift of his *Abba*. Hence, we can never speak of Jesus without *Abba* or of *Abba* without Jesus and the love between them.

---

26. Barr, "Abba isn't daddy," 28 – 4. As Barr points out, much exegetical and philological effort has been expended to explain the meaning of Abba. Opinions are divided between "emphatic" and "vocative" connotations. In my attempt to draw attention to the deep and tender intimacy between the Father and the Son, I tend to read Abba as an emphatic expression with a vocative function. Paul seems to make a similar point when he describes Christian experience (Rom 8:15 and Gal 4:6).

Let me in closing take you to the one place where we are allowed, in quivering silence, to peer even deeper into this mystery. In the moment of Jesus' death, when he yields up his spirit with a *"loud voice"* (Matt 27:50), we behold in this final moment how the Son in an ultimate act of filial surrender lets go of himself in a last shout of absolute giving. Allusions to Jesus' experience of divine abandonment aside, I see in this instance both *Father* and *Son* united in the unfathomable depth of trinitarian love, which gives the utmost and only in this way is able to affirm life even in death, the divine reconciliation of all things (Col 1:20).

This *Abba* is infinitely more concerned with the transformation of human beings than with us getting our formal theology right. After all, is it not true to say that in the figure of the Crucified every preconceived theological coherence is deconstructed, that in Jesus, crucified and risen, we are confronted with the incomprehensible reality that *Abba's* limitless compassion seeks us out, longing for it to become contagious among us? Given such a conception of the triune God, let us ponder in the next section what one might say about divine actions in the creation of an evolving cosmos.

Trinity

In our search for more understanding, the words of the Catalan theologian Raimundo Panikkar from half a century ago can point the way:

> The different traits here taken into account are brought together to form a tress which represents one of the deepest intuitions man has had and is still having from different points of view and with different names: the intuition of the threefold structure of reality, of the triadic oneness existing on all levels of consciousness and of reality, of the Trinity.[27]

We are not saying that the idea of the Trinity can be reduced to the discovery of a triple dimension of Being, nor that this aspect is a mere rational discovery. We are only affirming that the Trinity is the acme of a truth that permeates all realms of being and consciousness and that this vision links us together.

On this reading, the core doctrine of the Christian confession resonates well with the wider implications of the new cosmological picture of the world we have been considering throughout. Even at first sight, there

27. Panikkar, *Trinity and Religious Experience*, xi.

is a sense that the ground of the dynamic wholeness that undergirds and pervades all material reality may have something to do with the Trinity. A second look, using a wider aperture, will allow us to see why scholars of religion also discover intimations of the Trinity in non-Christian religious traditions, although differently expressed.[28]

Christians believe that God is eternally "triune" or three-in-one: Father–Son–Holy Spirit, whose self-subsisting unity is at the same time unity-in-diversity marked by a certain order, relationships, and precedence of operations. The Father originates, the Son reveals, and the Spirit executes, so that we can say that creation is the work of the Father through the Son by the Holy Spirit. From this triune and internally self-giving mystery the gifts of creation and grace flow in freedom and love.

Here "gift" is to be understood first as the reciprocal bestowal of love between the divine persons and their yielding to one another, revealing God as the truly living One. The internal dynamism of the Creator as Trinity is thus free from stagnation or passivity. The gift also manifests externally as the wellspring of creation, the divine energy as love. While God's love-based inner unity negates the possibility of any dualism at the heart of the universe, divine communication and love toward the creation bring forth an ordered cosmos of diversity.[29] By the same token, God's inner diversity makes room for creation's boundless variety, which gives rise to the widely diversified and beautiful forms of the natural world.

Moreover, this love energy renders all things, from quarks to quasars, alive and precious. Love also constitutes God's self-communication inward among the three divine persons (in an absolute sense) and outward in revelation (in a relative sense). Hence, a theological vision that unifies our understanding of the new cosmic story, from atoms to molecules, to ecosystems, even to consciousness and social structures, must somehow reflect such trinitarian traits as unitive living together characterized by mutual indwelling, loving interaction, and compassionate sharing. In other words, because God as Trinity is ultimate wholeness as well as the depth of dynamic love, God's self-communication in creation turns the unfolding of the cosmos into an ecstatic, intensely relational yet unitive display of his glory.

Held in the embrace of that love, all creatures are drawn forward toward their own becoming by the longing of love, for love without longing

---

28. Panikkar, *Trinity and Religious Experience*, 42.
29. Finlayson, "Trinity."

for unifying fulfillment is unthinkable. Hence, trinitarian love cannot be external to the creation, for what the Trinity brings forth springs from the depths of its own interiority. Put differently, since this love is mutual and reciprocal between divine subjects, its eternal flow cannot relate to the creation at a distance as one would love an object. For in the final analysis trinitarian love is creation's fundamental subject, which not only holds the diversity of creation in unitive wholeness but also imparts to every element its unique identity from within. Uncontainable by a finite creation, eternal love's overwhelming excess simultaneously serves to draw the creation ever onward toward greater depth, relationship, union, life, and becoming.

This conception modifies an important plank in the traditional understanding of creation, as Australian theologian Denis Edwards (1943–2019) has noted: God does not create "discrete individual beings through a series of interventions" but rather "in one divine act . . . [that] embraces the whole process . . . [enables] what is radically new to emerge in creation."[30]

If trinitarian love is indeed the matrix of creation enabling creation's evolutionary creativity, another plank in traditional ways of thinking of God in creation is profoundly challenged: to remain love in fullness and perfection, God can never overpower, coerce, or control. Instead, trinitarian love must act toward the creation outwardly exactly as this love moves internally within the Trinity. Such love opens its hands and grants the creation the freedom to be itself, risking, by the same token, even the possibility of being rejected by the creation without cause.[31]

Where then is God in evolution? The view we have just sketched speaks directly to this question. For it is God's love that provides the self-activating and novelty-generating energy toward the arrival of ever more relational and thus more complex organizations of the material world.

If we believe that God is pouring out his love as the matrix of the creative process, we are ready to grasp why God cannot act in creation with "poof power" like a magician who makes fully formed creatures. Instead, the three persons of the Trinity externalize in space-time what

---

30. Edwards, *God of Evolution*, 30–31, 76.

31. This point relates to one of the main reasons why I wrote this book. I wanted to inform and possibly persuade fellow travelers in our time to read books of this kind without the fear that these will undermine their faith, wanting them to see how the new world picture requires us to recast some of our traditional ideas and concepts of reality.

they do internally and personally in the reciprocal outpouring in the interior of their free Being. It seems therefore more fitting to speak of the creation as the ecstatic effect of Being-in-Love rather than in terms of causation. The effect on believers of this shift in perception is profound, for if the triune Being-in-Love expresses the glory of the Trinity in this way, they can relax even when the externalized reality takes forms that do not match their traditional conceptions of how creation came to be, say by way of evolution—that is, by creative-adaptive change over long periods. Such a conception also speaks to one of the most controversial subjects, the issue of God's wrath. Without attempting a comprehensive exposition, I find myself in agreement with New Testament scholar C. H. Dodd (1884–1973) that wrath does not describe "an attitude of God but the inevitable process of cause and effect in a moral universe."[32]

Lastly, in the Prologue of John's Gospel (John 1:1–18), Jesus Christ is put forward as the *Logos*, the universal and eternal principle of divine reason: "Through him all things were made and without him nothing was made that has been made" (John 1:3). According to the Prologue, this *Logos* became flesh in Jesus of Nazareth, regarded by the early church fathers as God's Son, who preexists from before all worlds and is presented in the Book of Revelation as "the faithful and true witness, the beginning of the creation of God" (Rev 3:14).

When the biblical writers present Jesus of Nazareth as the Son's bodily manifestation in history, they assert that in the contingency and singularity of this one life the ineffable arises from the depth of desolation, abandonment, and death in the *kenosis* of divine love as unfathomable compassion for the creation as the path of its redemptive transformation that transcends all contrariness, even on a cosmic scale.

World without End?—A Concluding Reflection

Traditionally, Christian theology thinks of creation in three dimensions: the original creation or God's activity in the beginning, ongoing or continuous creation, and new creation or the fulfillment of creation. However, by the time the sciences discovered cosmic and biological evolution, the latter two dimensions had been assimilated to the first, so that the term "creation" had come to mean cosmic beginnings in practice. Even

---

32. Dodd, *Epistle to the Romans*, 23.

today in discussions between scientists and theologians, "creation" is often confined to the events surrounding the big bang.

However, such a limited conception of creation seriously divests this great divine project of its depth and ultimate outcome, namely, the participation of all creatures in the life of the Creator. "Creation," then, refers to a much more comprehensive concept that signifies the single yet inseparably tripartite project: the original creation, its redemption, and the new creation, whereby the Creator acts in radically new ways to bring about each of these three dimensions. In the original creation, God—in freedom and love—energizes a self-creating and incomplete universe as cocreator enabled to bring forth creatures of increasing complexity. In redemption, God reveals to the fullest the love-nature of his project and demonstrates proleptically its ultimate destiny through the selfless participation of Jesus of Nazareth in life, death, and resurrection. In the new creation, God will bring the divine project to ultimate completion so that the "creation will be unambiguously good."[33] The new creation is not a second creation from nothing (*ex nihilo*) but the transformation of the creation such that "the suffering of creatures is taken up *individually* by God in the Incarnation, suffered by God in the crucifixion, and redeemed by God in the Resurrection."[34]

To reiterate, creation so understood is one single divine project that includes from the beginning an undeniably transformational orientation toward God's ultimate end. This proposal takes the discovery of evolution, both cosmic and biological, seriously; indeed, it regards it as a gift for it allows us to appreciate more fully that *creation* means more than the origin of the cosmos. Moreover, it invites us to realize that the original creation could never have been instantly complete and perfect at the start, as millions of Christians believe. Our trinitarian reflections also led us to see that this ongoing creation was not driven from the past but in its unfolding was being addressed by God from the future. One might even say that cosmic and biological evolution are built-in creative modes of nature's response to the coming of God that ensure the intrinsic lawfulness and freedom of the creation to experiment adaptively with new forms in its contingent and unfinished state.

However, these considerations notwithstanding, significant difficulties arise when we bring the Christian doctrine of ultimate hope

---

33. Russell, "Eschatological Hope."
34. Russell, "Eschatological Hope," 106.

(eschatology) into dialogue with the natural sciences, especially the predictions of big bang cosmology. Let me touch on some of the issues involved.

When Christianity came into being, it was from the beginning a religion of cosmic hope. The infant church expected the immanent return of Christ and proclaimed the end of the present order. The old order was to be replaced by the soon-to-be-realized arrival of a radically transformed world of which the resurrection of Jesus was and would remain the defining event. With the rise of modern cosmology during the second half of the twentieth century, two German theologians, Wolfhart Pannenberg (1928–2014) and Jürgen Moltmann (1926–), began to embrace the new picture of the universe in their theology of hope, without developing a creative interaction with the natural sciences. Another towering figure among this group is the American Lutheran theologian Ted Peters (1941–). In the ensuing decades, this task was taken up by several scientist-theologians, including Arthur Peacocke (1924–2006), John Polkinghorne (1930–2021), and Robert John Russell (1946–), whose brilliant and unparalleled contributions to the science-theology dialogue cannot be overstated.

The magnitude of the scientific challenge is thrown into sharp relief with a simple question: Assuming no catastrophic events intervene, like a nuclear holocaust or an asteroid impact, how long can the Earth support life? The short answer is that all life on Earth will end in 1.75–3.25 billion years, when the planet is driven out of the habitable zone. Then the sun will begin to inflate due to diminishing hydrogen fuel as it turns into a red giant star whose circumference will eventually exceed Earth's orbit. Long before the sun swallows Earth, the oceans will have evaporated, and Earth's surface temperature will have risen beyond the endurance of the most resilient microbes. Needless to say, human life will have ceased hundreds of millions of years earlier.

Given the geophysical dynamics of our planet, other events will have made their influence felt as well. Approximately 250–350 million years from now, tectonic plate movements will create a new supercontinent with unpredictable effects on the Earth's axial tilt, possibly altering the magnetic field that protects life from deadly cosmic radiation. Six hundred million years from now, when increasing sun radiation will have depleted atmospheric molecules of carbon dioxide, all photosynthesis will cease along with all plant life.

On a larger scale and on current predictions, the universe itself will continue its trajectory of expansion for more than 10 billion years, depending on the shape of the two main drivers: its geometry and its dark energy content. In other words, if general relativity is correct, the final state of the universe may best be described by the well-worn adage "freeze or fry." Ted Peters sums up the seriousness of the scientific challenge for Christianity with audacious honesty: "Should the final future of the combination of big bang cosmology and the second law of thermodynamics come to pass . . . then we would have proof that our faith has been in vain. It would turn out that there is no God, at least not the God in whom followers of Jesus have put their trust [although no living creature would be around to witness let alone report it]."[35] A dire prediction indeed! It represents the high point of reductionist scientific materialism, which proclaims a universe made up of nothing but dead and mindless matter.

Since such a universe—even if it lasted for trillions of years—would be utterly pointless, this view epitomizes the fatalism associated with a materialist conception of the world that sullies every creaturely aspiration toward self-transcendence with hopelessness and futility. One would even question whether striving toward goodness, beauty, and self-sacrificial love could even arise in such a cosmos, leaving us with all-out hedonism as the only sane response. However, since universal human experience suggests otherwise, we are prompted to ask another question: Where is true hope to be found, the kind of hope that does not buckle under the threat of cosmic absurdity? Exploring this question—in particular, at the end of our reflections on the cosmos as God's creation in the age of science—seems to offer a fitting closure.

In the preceding chapters I have referred in several places to the enormous differences between what modern cosmology tells us about the cosmos and the traditional biblical narrative. Let me add one more such feature: from the standpoint of science, it is not inescapably obvious that there should only be one universe that currently evolves according to the laws we observe, to the exclusion of other universes based on totally different laws. This possibility takes the conflict between modern cosmology and the doctrine of Christian hope or eschatology to another level. After all, science has no interest in Christian claims that promise a redemptive transformation and renewal of the cosmos. Its own long-term projections are derived from straight-line extrapolations of currently

---

35. Peters, quoted in Russell, "Eschatology and Cosmology," 284.

discernible and mathematically describable patterns of observation into a far future measured in tens of billions of years. By contrast, the church expects a future disruption patterned on the resurrection of Jesus or the inbreaking of a radical discontinuity called the kingdom of God into the creation (as it may exist at the time of this event), showing up the obvious: the concerns of science and of Christian eschatology seem to be fundamentally opposed to each other. The natural sciences account for novelty in the universe through emergence over time, while Christian hope is grounded in the *radical transformation* that God will bring about in the future when he acts in new ways.[36]

On the surface, these two rival paradigms seem to have nothing in common, one being based on observable data and derivative theories, the other on the freedom of God and the biblical promise of "a new heaven and a new earth" (Rev 21:1–4). Scientist-theologian Robert John Russell is convinced that "creative mutual interaction" between them is nonetheless possible. In response to this challenge, Russell has conducted an ambitious research program.[37] His life's work, always in conversation with his peers, allows us to identify three theological claims. There exists:

- A contingent universe whose laws are the result of God's creative nature and intent; they describe how the universe works while testifying to God's faithfulness and constancy.

- The biblical testimony, which tells us that God works differently in different epochs: it empowered the first Easter community to understand Jesus' crucifixion and resurrection as the beginning of the end of the world and the dawn of God's kingdom.

- A future when we can expect God to act again to bring the creation to its ultimate fulfillment. Indeed, the church believes that this final stage has already begun in an anticipatory way with the resurrection of Jesus as elaborated in Pannenberg's notion of "prolepsis."

On this view, the whole cosmos is moving toward its ultimate destiny, energized by trinitarian love and the divine promise that one day "God will be all in all" (2 Cor 15:28). While time lasts, Christian expectancy is directed toward the participation of the entire creation in the life of the Trinity and the deification of individuals through the divine Spirit. While

36. Russell, "Eschatology and Cosmology," 282.

37. Russell, *Cosmology, Evolution, and Resurrection*; also, Russell, *Cosmology: From Alpha to Omega*.

destined to embody divine love as bearers of the divine image (Gen 1:27), believers recapitulate cosmic history, including creation's inward groaning, until it is accomplished (Rom 8:21–22). Or, in another way, with the eyes of hindsight and anticipation we behold God's magnificent project of creation, redemption, and new creation as a movement toward a new future where love will reign supreme.

The challenge for Christianity remains whether it can, at this crucial stage of history, fuse evolution with divine action into a single creative-theological comprehension that embraces the entire cosmos. In my view, it will involve grasping Haught's insight that with the rise of "human intelligence, moral aspiration, aesthetic sensibility, and religious wonder" the cosmos itself is awakening to consciousness, and that this awakening includes the understanding that "the blinding sun of rightness" has yet to fully appear above the horizon in an unfinished creation.[38]

To do justice to its first flaring forth almost 14 billion years ago out of nowhere, to its dramatic unfolding during its long history, to its life-generating power and its awakening to sentience, Christians need to understand this universe, including the sciences that describe it, as a glorious yet incomplete narrative. Certainly, to this day its unsearchable depth and wholeness remain hidden. Yet when uncontainable trinitarian love broke through the limits of human comprehension with the life, death, and resurrection of Jesus, something new was articulated: Christ's love has appeared as cosmic inwardness, offering us humans at this moment in history a part in the discovery and further unfolding of that love and inviting us to orient ourselves toward its open horizon with reckless abandon. Thus, Abba's gratuitous triune love draws our hearts and minds far beyond earlier conceptions of the natural world and of its creator that we might become more fully those through whom this love lives.

---

38. Haught, *New Cosmic Story*, 200.

Bibliography

Alison, James. *The Joy of Being Wrong: Original Sin through Easter Eyes.* New York: Crossroad, 1998.

Allen, Douglas. "Mircea Eliade's Phenomenological Analysis of Religious Experience." *Journal of Religion* 52/2 (April 1972) 170–186.

"Animal Consciousness." *Stanford Encyclopedia of Philosophy.* 2016. https://stanford.library.sydney.edu.au/entries/ consciousness-animal/index.html.

Aquinas, Thomas. *Summa Theologiae.* Vol. 54. Edited by Richard T. A. Murphy. New York: McGraw-Hill, 1963.

———. *Summa Theologiae.* Literally translated by fathers of the English Dominican Province. 2nd ed. Originally published 1920. Online ed., 2017 by Kevin Knight. https://www.newadvent.org/summa/1.htm.

Aslan, Reza. *God: A Human History.* New York: Random House, 2017.

"Atapuerca Archaeological Site." UNESCO World Heritage List. http://whc.unesco.org/en/list/989.

Augustine of Hippo. *On Christian Doctrine.* Translated by James Shaw. In *Nicene and Post-Nicene Fathers,* 1st ser., edited by Philip Schaff, vol. 2. Buffalo, NY: Christian Literature, 1887. Revised and edited for New Advent by Kevin Knight. https://www.newadvent.org/fathers/1202.htm.

———. *City of God.* Translated by Marcus Dods. In *Nicene and Post-Nicene Fathers,* 1st ser., edited by Philip Schaff, vol. 2. Buffalo, NY: Christian Literature, 1887. Revised and edited for New Advent by Kevin Knight. https://www.newadvent.org/fathers/1201.htm.

"Australia's Warratyi Rock Shelter Excavated." *Archaeology,* November 3, 2016. https://www.archaeology.org/news/4990-161103-australia-megafauna-rock-shelter.

Ayala, Francisco J. "Biological Evolution." In *An Evolving Dialogue: Theology and Scientific Perspectives on Evolution,* edited by James B. Miller, 9–52. Harrisburg, PA: Trinity, 2001.

Bader, Christopher, et al. *American Piety in the 21st Century: New Insights to the Depth and Complexity of Religion in the US.* Waco, TX: Baylor University Press, 2006. https://www.baylor.edu/content/services/document.php/33304.pdf.

Bailie, Gil. *Violence Unveiled: Humanity at the Crossroads.* New York: Crossroad, 1995.

Barfield, Owen. *Saving the Appearances: A Study in Idolatry.* 2nd ed. Middletown, CT: Wesleyan University Press, [1957] 1988.

Barrow, John D., and Frank J. Tipler. *The Anthropic Cosmological Principle.* Oxford: Oxford University Press, 1986.

Barr, James. "Abba isn't daddy." Journal of Theological Studies 39 (1988) 28–47.
Bartlett, Anthony W. *Cross Purposes: The Violent Grammar of Christian Atonement*. Harrisburg, PA: Trinity, 2001.
Basil of Caesarea. *Hexaemeron*. Translated by Blomfield Jackson. In *Nicene and Post-Nicene Fathers*, 2nd ser., edited by Philip Schaff and Henry Wace, vol. 8. Buffalo, NY: Christian Literature, 1895. Revised and edited for New Advent by Kevin Knight. htpps://newadvent.org/fathers/cathen/02330b.htm.
Baudler, Georg. *Das Kreuz: Geschichte und Bedeutung*. Düsseldorf: Patmos, 1997.
———. *Die Befreiung von einem Gott der Gewalt: Erlösung in der Religionsgeschichte von Judentum, Christentum und Islam*. Düsseldorf: Patmos, 1999.
———. *God and Violence: The Christian Experience of God in Dialogue with Myths and Other Religions*. Translated by Fabian C. Lochner. Springfield, IL: Templegate, 1992.
Biser, Eugen. *Der Mensch: Das Uneingelöste Versprechen*. Düsseldorf: Patmos, 1996.
———"Die Stimme der Antigone: zu Georg Baudlers Untersuchung von Gewalt und Gewaltlosigkeit in Religion und Christentum." *Theologische Revue* 50/9 (1994) 355–64, 367–68.
Bohm, David. *Wholeness and the Implicate Order*. London: Ark, [1980] 1988.
Bor, Daniel. *The Ravenous Brain: How the New Science of Consciousness Explains Our Insatiable Search for Meaning*. New York: Basic, 2012.
Braterman, Paul. "'God Intended It as a Disposable Planet': Meet the US Pastor Preaching Climate Change Denial." *The Conversation*, October 12, 2020. https://theconversation.com/god-intended-it-as-a-disposable-planet-meet-the-us-pastor-preaching-climate-change-denial-147712.
Brecht, Martin. *Martin Luther*, vol. 3: *The Preservation of the Church, 1532–1546*. Minneapolis: Fortress, 1993.
Brueggemann, Walter. *An Introduction to the Old Testament: The Canon and Christian Imagination*. Louisville: Westminster John Knox, 2003.
Brunner, Emil. *Truth as Encounter*. 2nd ed. London: SCM, 1964.
Burkert, Walter. *Homo Necans: The Anthropology of Ancient Greek Sacrificial Ritual and Myth*. Berkeley: University of California Press, 1983.
"Carl Sagan." *Wikipedia*, 2021. https://en.wikipedia.org/wiki/Carl_Sagan.
Carroll, Sean B. *Endless Forms Most Beautiful: The New Science of Evo Devo*. New York: Norton, 2005.
Catechism of the Catholic Church. Chicago: Loyola University Press, 1994.
"Child Sacrifice in Uganda." *Wikipedia*, 2021. https://en.wikipedia.org/wiki/Child_sacrifice_in_Uganda.
Chouraqui, André. *Entête (La Genèse)*. Paris: Jean-Claude Lattès, 1992.
Christian, David. *Maps of Time: An Introduction to Big History*. 2nd ed. Berkeley: University of California Press, 2011.
"Church Attendance." *Wikipedia*, 2021. https://en.wikipedia.org/wiki/Church_attendance.
Clement of Alexandria. *The Stromata or Miscellanies*. Translated by William Wilson. In *Ante-Nicene Fathers*, edited by Alexander Roberts, James Donaldson, and A. Cleveland Coxe, vol. 2. Buffalo, NY: Christian Literature, 1885. Revised and edited for New Advent by Kevin Knight. http://www.newadvent.org/fathers/0210.htm.
Cowell, Siôn. *The Teilhard Lexicon: Understanding the Language, Terminology and Vision of the Writings of Pierre Teilhard de Chardin*. Brighton, UK: Sussex Academy, 2001.

Darwin, Charles. *On the Origin of Species*. London: John Murray, 1859.
Daum, Werner. *Ursemitische Religion*. Stuttgart: Kohlhammer, 1985.
Davies, Paul. *The Goldilocks Enigma: Why the Universe Is Right for Life*. London: Allen Lane, 2006.
Delio, Ilia. *Christ in Evolution*. Maryknoll, NY: Orbis, 2008.
———. *The Unbearable Wholeness of Being: God, Evolution, and the Power of Love*. Maryknoll, NY: Orbis, 2016.
Dendrinos, Dimitrios S. "Dating Gobekli Tepe: The Evidence Doesn't Support a PPNB Date, but Instead a Possibly Much Later One." September 19, 2016. https://www.researchgate.net/publication/317433791_Dating_Gobekli_Tepe.
Dobzhansky, Theodosius. *Mankind Evolving: The Evolution of the Human Species*. New Haven, CT: Yale University Press, 1962.
Dodd, Charles Harold. *The Epistle of Paul to the Romans*. London: Hodder and Stoughton, 1932.
d'Olivet, Fabre. *La Langue Hébraïque Restituée*. Paris: Éditions de la Tête de Feuille, 1971.
Donald, Merlin. "Mimesis Theory Re-examined Twenty Years after the Fact." In *Evolution of Mind, Brain, and Culture*, edited by Gary Hatfield and Holly Pittman, 169–92. Philadelphia: University of Pennsylvania Museum of Archaeology and Anthropology, 2013.
Donaldson, Margaret. *Children's Minds*. Rev. ed. New York: Norton, 1979.
Douglas-Klotz, Neil. *The Hidden Gospel: Decoding the Spiritual Message of the Aramaic Jesus*. Wheaton, IL: Theosophical, 2001.
Dumouchel, Paul, ed. *Violence and Truth: On the Work of René Girard*. Stanford, CA: Stanford University Press, 1988.
Du Troit, Cornel W. "Towards a New Natural Theology Based on Horizontal Transcendence." *HTS Theological Studies* 65/1 (April 2009) 243–50.
Edwards, Denis. *The God of Evolution: A Trinitarian Theology*. Mahwah, NJ: Paulist, 1999.
Eliade, Mircea. *Mephistopheles and the Androgyne*. Translated by J. M. Cohen. New York: Sheed & Ward, 1965.
———. *Myths, Dreams, and Mysteries*. Translated by Philip Mairet. New York: Harper & Row, 1960.
———. *Patterns in Comparative Religion*. Translated by Rosemary Sheed. New York: Meridian, 1966.
———. *The Sacred and the Profane*. Translated by Willard R. Trask. New York: Harper, 1961.
———. "Structure and Changes in the History of Religion." Translated by Kathryn Atwater. In *Invincible City: A Symposium on Urbanization and Cultural Development in the Ancient Near East Held at the Oriental Institute of the University of Chicago, December 4–7, 1958*, edited by Carl H. Kraeling and Robert McC. Adams, 351–65. Chicago: University of Chicago Press, 1960.
Ellul, Jacques. "Innocent Notes on 'The Hermeneutic Question.'" In *Jacques Ellul, Sources and Trajectories: Eight Early Articles by Jacques Ellul That Set the Stage*, edited and translated by Marva J. Dawn, 184–203. Grand Rapids: Eerdmans, 1968.
———. *On Freedom, Love, and Power*. Edited and translated by Willem H. Vanderburg. Toronto: University of Toronto Press, 2010.

———. *Sources and Trajectories: Eight Early Articles by Jacques Ellul That Set the Stage*. Edited and translated by Marva J. Dawn. Grand Rapids: Eerdmans, 1997.

Erickson, Millard. *Christian Theology*. Grand Rapids: Baker Book House, 1985.

Evans, C. Stephen. *Natural Signs and Knowledge of God*. Oxford: Oxford University Press, 2010.

Finlan, Stephen. *Problems with Atonement: The Origins of, and Controversy about, the Atonement Doctrine*. Collegeville, MN: Liturgical, 2005.

Finlayson, R. A. "Trinity." In *New Bible Dictionary*, edited by I. D. Douglas et al., 1298–1300. Leicester, UK: InterVarsity, 1962.

Fleming, Chris. *René Girard: Violence and Mimesis*. Cambridge: Polity, 2004.

Frank, Adam. *The Constant Fire: Beyond the Science vs. Religion Debate*. Berkeley: University of California Press, 2009.

Funes, José G., Lucio Florio, Marcelo Lares, and Mariano Asia. "Searching for Spiritual Signatures in SETI Research." *Theology and Science* 17/3 (2019) 373–81.

Garrels, Scott R. "Human Imitation: Historical, Philosophical, and Scientific Perspectives." In *Mimesis and Science: Empirical Research on Imitation and the Mimetic Theory of Culture and Religion*, edited by Scott R. Garrels, 1–38. East Lansing: Michigan State University Press, 2011.

———. "Mimesis and Science: An Interview with René Girard." In *Mimesis and Science: Empirical Research on Imitation and the Mimetic Theory of Culture and Religion*, edited by Scott R. Garrels, 215–54. East Lansing: Michigan State University Press, 2011.

Giegerich, Wolfgang. *Tötungen, Gewalt aus der Seele: Versuch über Ursprung und Geschichte des Bewusstseins*. Frankfurt: Peter Lang, 1994.

Girard, René. *Battling to the End: Conversations with Beno[circumflex over i]ît Chantre*. Translated by Mary Baker and Andrew McKenna. East Lansing: Michigan State University Press, 2010.

———. *Sacrifice*. Translated by Matthew Pattillo and David Dawson. Breakthroughs in Mimetic Theory. East Lansing: Michigan State University Press, 2011.

———. *Violence and the Sacred*. Translated by Patrick Gregory. Baltimore: Johns Hopkins University Press, 1977.

Gnuse, Robert K. "The Emergence of Monotheism in Ancient Israel: A Survey of Recent Scholarship." *Religion* 29 (1999) 319–23.

Goldstein, Jeffrey. "Emergence as a Construct: History and Issues." *Emergence: Complexity and Organization* 1/1 (1999) 49–72.

Goodall, Jane. *The Chimpanzees of Gombe: Patterns of Behavior*. Cambridge, MA: Belknap, 1986.

Greenspan, Stanley I., and Stuart G. Shanker. *The First Idea: How Symbols, Language, and Intelligence Evolved from Our Primate Ancestors to Modern Humans*. Cambridge, MA: Da Capo, 2004.

Gregory the Great. *Moralia*. In *A History of Christian Thought*, by Arthur Cushman McGiffert, vol. 2. New York: Scribner, 1933.

Gutmann, H.-M. *Die tödlichen Spiele der Erwachsenen: Moderne Opfermythen in Religion, Politik und Kultur*. Freiburg: Herder, 1995.

Hameroff, Stuart R. "Funda-Mentality: Is the Conscious Mind Subtly Linked to a Basic Level of the Universe?" https://quantumconsciousness.org/content/funda-mentality.

Hameroff, Stuart R., and Roger Penrose. "Orchestrated Reduction of Quantum Coherence in Brain Microtubules: A Model for Consciousness." In *Toward a Science of Consciousness: The First Tucson Discussions and Debates*, edited by Stuart R. Hameroff, Alfred W. Kaszniak, and Alwyn C. Scott, 507–40. Cambridge, MA: MIT Press, 1996.

Hamerton-Kelly, Robert. "Religion and the Thought of René Girard." In *Curing Violence: Essays on René Girard*, edited by Mark I. Wallace and Theophus H. Smith, 3–24. Sonoma, CA: Polebridge, 1994.

Harlan, Chico. "The Fire Came at a Difficult Time for French Catholics." *Washington Post*, April 17, 2019. https://www.washingtonpost.com/world/europe/notre-dame-fire-came-at-a-difficult-time-for-french-catholics/2019/04/17/.

Hatfield, Gary. "Introduction: The Evolution of Mind, Brain, and Culture." In *The Evolution of Mind, Brain, and Culture*, edited by Gary Hatfield and Holly Pittman, 1–44. Philadelphia: University of Pennsylvania Museum of Archaeology and Anthropology, 2013.

Haught, John F. *Deeper than Darwin: The Prospect of Religion in the Age of Evolution*. Boulder, CO: Westview, 2003.

———. *God after Darwin: A Theology of Evolution*. Boulder, CO: Westview, 2000.

———. *The New Cosmic Story: Inside Our Awakening Universe*. New Haven, CT: Yale University Press, 2017.

———. *The Revelation of God in History*. Wilmington, DE: Michael Glazier, 1988.

Hawking, Stephen. *A Brief History of Time*. New York: Bantam, 1988.

Hay, David. *Something There: The Biology of the Human Spirit*. Philadelphia: Templeton Foundation, 2007.

Head, Ivan. "Herodian Aspects of the English Reformation Monarchy: Anglican Brutality and Girardian Insight." In *Violence, Desire, and the Sacred*, edited by Scott Cowdell, Chris Fleming, and Joel Hodge, vol. 1, *Girard's Mimetic Theory across the Disciplines*, 182–205. New York: Bloomsbury, 2012.

Heathcote, Angela. "Tawny Frogmouths: Five Things You May Not Know About These Masters of Disguise," Australian Geographic, May 16 (2018). https://www.australiangeographic.com.au/topics/wildlife/2018/05/tawny-frogmouths-5-things-you-may-not-know-about-these-masters-of-disguise/

Henshilwood, Christopher S., et al. "An Early Bone Tool Industry from the Middle Stone Age at Blombos Cave, South Africa: Implications for the Origins of Modern Human Behaviour, Symbolism and Language." *Journal of Human Evolution* 41 (2001) 631–78.

Herzog, Markwart. "Religionstheorie und Theologie René Girards." *Kerygma und Dogma* 38/2 (April/June 1992) 105–37.

Hillman, Anne. *Awakening the Energies of Love: Discovering Fire for the Second Time*. North Bergen, NJ: Bramble, 2008.

Hoffman, Joel M. *And God Said: How Translations Conceal the Bible's Meaning*. Thomas Dunne, 2010.

Horner, Robyn. "Words That Reveal: Jean-Yves Lacoste and the Experience of God." *Continental Philosophy Review*, May 28, 2017. DOI 10.1007/S11007-017-9420-x.

James, Steven R. "Hominid Use of Fire in the Lower and Middle Pleistocene: A Review of the Evidence." *Current Anthropology* 30/1 (February 1989) 1–26.

Janssen, Luke J. *Standing on the Shoulders of Giants: Genesis and Human Origins*. Eugene, OR: Wipf & Stock, 2016.

Jeans, James. *The Mysterious Universe.* New York: Macmillan, 1931.

Johnson, Elizabeth. "Presidential Address: Turn to the Heavens and the Earth: Retrieval of the Cosmos in Theology." *CTSA Proceedings* 51 (1996) 1–14.

Jonas, Hans. *Mortality and Morality: A Search for the Good after Auschwitz.* Edited by Lawrence Vogel. Evanston, IL: Northwestern University Press, 1996.

Kauffman, Stuart A. *Reinventing the Sacred: A New View of Science, Reason, and Religion.* New York: Basic Books, 2008.

Kirk-Duggan, Cheryl A. "Gender, Violence, and Transformation in Alice Walker's *The Color Purple*." In *Curing Violence: Essays on René Girard*, edited by Mark I. Wallace and Theophus H. Smith, 266–86. Sonoma, CA: Polebridge, 1994.

Kirwan, Michael. *Discovering Girard.* London: Darton, Longman, & Todd, 2004.

König, Marie E. P. *Am Anfang der Kultur: Die Zeichensprache des frühen Menschen.* Berlin: Gebr. Mann, 1973.

Langer, Susanne K. *Philosophy in a New Key: A Study in the Symbolism of Reason, Rite, and Art.* Cambridge, MA: Harvard University Press, 1942.

Lee, T. F., F. Mora, and R. D. Myers. "Dopamine and Thermoregulation: An Evaluation with Special Reference to Dopaminergic Pathways." *Neuroscience & Biobehavioral Reviews* 9/4 (Winter 1985) 589–98.

Leidenhag, Joanna. "The Revival of Panpsychism and Its Relevance for the Science-Religion Dialogue." *Theology and Science* 17/1 (2019) 90–106.

Lewis, C. S. *Surprised by Joy: The Shape of My Early Life.* San Francisco: HarperOne, [1955] 2017.

Lewis-Williams, David J., and Jean Clottes. "The Mind in the Cave—the Cave in the Mind: Altered States of Consciousness in the Upper Paleolithic." *Anthropology of Consciousness* 9/1 (1998) 13–21.

Lockerby, Patrick. "Henry Ford – Quote: 'History Is Bunk.'" *Science 2.0*, The Chatter Box, May 30, 2011. https://www.science20.com/chatter_box/henry_ford_quote_history_bunk-79505.

Lommel, Hermann. *Die Religion Zarathustras: Nach dem Awesta Dargestellt.* Hildesheim: Olms, 1971.

Lovell, Bernard. *In the Centre of Immensities.* World Perspectives 53. London: Book Club Associates, by arrangement with Hutchinson and Co., 1979.

Lücke, Ulrich. "Der Mensch—das Religiöse Wesen aus Ethologischer, Paläoanthropologischer und Theologischer Perspektive." In *Religion: Entstehung, Funktion, Wesen*, edited by Hans Waldenfels, 71–92. Freiburg: Karl Alber, 2002.

Luther, Martin. *On the Jews and Their Lies.* Translated by Martin H. Bertram. In *Luther's Works*, vol. 47, edited by Robert C. Schultz, 268–71. Philadelphia: Fortress, 1971.

Mack, Burton L. *The Christian Myth: Origins, Logic, and Legacy.* New York: Continuum, 2001.

———. "The Innocent Transgressor: Jesus in Early Christian Myth and History." *Semeia* 33 (1985) 135–65.

McHenry, Henry. "Homo heidelbergensis." *Encyclopedia Britannica*, June 19, 2020. https://www.britannica.com/topic/Homo-heidelbergensis.

Meltzoff, Andrew, and Keith Moore, "Imitation of Facial and Manual Gestures by Human Neonates." *Science* 198 (1977) 75–78.

Moltmann, Jürgen. *The Crucified God: The Cross of Christ as the Foundation and Criticism of Christian Theology.* Translated by R. A. Wilson. London: SCM, 1974.

———. *Theology of Hope.* Translated by James W. Leitch. London: SCM, 1967.

Moore, Sebastian. "Foreword." In James Alison, *The Joy of Being Wrong: Original Sin through Easter Eyes*, vii–xi. New York: Crossroad, 1998.
"Neanderthals Having Disabilities Survived with the Help of Social Support." *The TeCake*, October 25, 2017. https://tecake.in/neanderthals-disabilities-survived-help-social support.
Negev, Avraham, and Shimon Gibson. "Arad (Tel)." In *Archaeological Encyclopedia of the Holy Land*, 43. New York: Continuum, 2001.
Newberg, Andrew. *The Spiritual Brain: Science and Religious Experience*. Chantilly, VA: Teaching Company, 2012.
Nietzsche, Friedrich. *Also sprach Zarathustra*. Werke: Kritische Gesamtausgabe 1.6. Berlin: Walter de Gruyter, 1968.
Otto, Rudolf. *The Idea of the Holy: An Inquiry into the Non-Rational Factor in the Idea of the Divine and Its Relation to the Rational*. Translated by John W. Harvey. London: Oxford University Press, 1923.
Oughourlian, Jean-Michel. "From Universal Mimesis to the Self Formed by Desire." In *Mimesis and Science: Empirical Research on Imitation and the Mimetic Theory of Culture and Religion*, edited by Scott R. Garrels, 41–54. East Lansing: Michigan State University Press, 2011.
Packer, J. I. "Revelation." In *New Bible Dictionary*, edited by J. D. Douglas et al., 1090–93. Rev. ed. Leicester, UK: InterVarsity, 1977.
Palaver, Wolfgang. *René Girard's Mimetic Theory*. Translated by Gabriel Borrud. East Lansing: Michigan State University Press, 2013.
Panikkar, Raimundo. *The Trinity and the Religious Experience of Man*. 2nd ed. London: Darton, Longman, & Todd, 1973.
Papachristophorou, Marilena. "Ordinary Stories, Dreams, Miracles and Social Interactions." *Journal of Comparative Research in Anthropology and Sociology* 5/2 (Winter 2014) 61–71.
Peacocke, Arthur R. *The Palace of Glory: God's World and Science*. Adelaide: Australian Theological Forum, 2005.
Penrose, Roger, and Stuart R. Hameroff. "What 'Gaps'? Reply to Grush and Churchland." *Journal of Consciousness Studies* 2/2 (1995) 99–112.
Pew Research Center. *Being Christian in Western Europe*. Pew Research Center, May 29, 2018. https://www.pewforum.org/2018/05/29/being-christian-in-western-europe/.
Plato. *Timaeus*. Translated by R. G. Bury. Loeb Classical Library. Cambridge, MA: Harvard University Press, 1929.
Polanyi, Michael. *Personal Knowledge: Towards a Post-Critical Philosophy*. New York: Harper, 1958.
Polkinghorne, John. *One World: The Interaction of Science and Theology*. London: SPCK, 1986.
———. *Beyond Science: The Wider Human Context*. Cambridge: Cambridge University Press, 1996.
Previc, Fred H. "Dopamine and the Origins of Human Intelligence." *Brain and Cognition* 41/3 (2000) 299–350.
———. *The Dopaminergic Mind in Evolution and History*. Cambridge: Cambridge University Press, 2009.
Primack, Joel R., and Nancy Ellen Abrams. *The View from the Center of the Universe: Discovering Our Extraordinary Place in the Cosmos*. New York: Riverhead, 2006.

Ramsey, Ian T. *Religious Language: An Empirical Placing of Theological Phrases.* London: SCM, 1969.

Rashdall, Hastings. *The Idea of Atonement in Christian Theology.* London: Macmillan, 1919.

Rightmire, G. Philip. "Homo habilis." *Encyclopedia Britannica,* November 12, 2020. https://www.britannica.com/topic/Homo-habilis.

"Ritual Killing and Human Sacrifice in Africa." *Humanists International,* November 12, 2010. https://humanists.international/20011/10/ritual-killing-and-human-sacrifice-africa.

Rohr, Richard. *Everything Belongs: The Gift of Contemplative Prayer.* Rev. ed. New York: Crossroad, 2003.

Rovelli, Carlo. *Seven Brief Lessons on Physics.* Translated by Simon Carnell and Erica Segre. New York: Riverhead /Penguin Random House, 2016.

Russell, Heidi Ann. *Quantum Shift: Theological and Pastoral Implications of Contemporary Developments in Science.* Ebook. Collegeville, MN: Liturgical, 2015.

———. "Sanctity of Science: The Mysticism of Theologically Engaging Science." *Theology and Science* 10/3 (2012) 249–58.

Russell, Robert John. *Cosmology: From Alpha to Omega.* Minneapolis: Fortress, 2008.

———. *Cosmology, Evolution, and Resurrection Hope: Theology and Science in Creative Mutual Interaction.* Edited by Carl S. Helrich. London: Pandora, 2006.

———. "Eschatological Hope: A Grateful Response to Tom Tracy." *Theology and Science* 10/1 (February 2012) 103–6.

———. "Eschatology and Physical Cosmology: A Preliminary Reflection." In *The Far-Future Universe: Eschatology from a Cosmic Perspective,* edited by George F. R. Ellis, 266–315. Philadelphia: Templeton, 2002.

———. *Time in Eternity: Pannenberg, Physics, and Eschatology in Creative Mutual Interaction.* Notre Dame, IN: University of Notre Dame Press, 2012.

Sagan, Carl. *The Dragons of Eden: Speculations on the Evolution of Human Intelligence.* New York: Ballantine, 1977.

———. *Pale Blue Dot: A Vision of Human Future in Space.* New York: Ballantine, 1997.

Schroeder, Gerald L. *The Hidden Face of God: Science Reveals the Ultimate Truth.* New York: Free, 2001.

Schwager, Raymund. *Banished from Eden: Original Sin and Evolutionary Theory in the Drama of Salvation.* Translated by James Williams. Leominster, Herefordshire, UK: Gracewing, 2006.

———. *Must There Be Scapegoats?* San Francisco: Harper & Row, 1987.

Shurr, Theodore G. "When Did We Become Human? Evolutionary Perspectives on the Emergence of the Modern Human Mind, Brain, and Culture." In *Evolution of Mind, Brain, and Culture,* edited by Gary Hatfield and Holly Pittman, 45–89. Philadelphia: University of Pennsylvania Museum of Archaeology and Anthropology, 2013.

Smithsonian. "Mating in Insects." https://www.si.edu/spotlight/buginfo/mating.

Smithsonian National Museum of Natural History, Human Origins Initiative. "The Oldest Fossil of Our Genus." March 5, 2015. https://humanorigins.si.edu/research/whats-hot-human-origins/oldest-fossil-our-genus.

Southgate, Christopher. *The Groaning of Creation: God, Evolution, and the Problem of Evil.* Louisville: Westminster John Knox, 2008.

———. "Science and Religion in the United Kingdom: A Personal View on the Contemporary Scene." *Zygon* 51/2 (June 2016) 361–86.

Sporns, Olaf. "Network Analysis, Complexity and Brain Function." *Complexity* 8/1 (2003) 56–60.

Starr, Michelle. "Study Maps the Odd Structural Similarities between the Human Brain and the Universe." *Science Alert*, November 17, 2020. https://www.sciencealert.com/wildly-fun-new-paper-compares-the-human-brain-to-the-structure-of-the-universe.

Swaminathan, Nikhil. "Why Does the Human Brain Need So Much Power?" *Scientific American*, April 29, 2008. https://www.scientificamerican.com/article/why-does-the-brain-need-s/.

Swimme, Brian Thomas, and Mary Evelyn Tucker. *Journey of the Universe*. New Haven, CT: Yale University Press, 2011.

Tanner, Nancy M. *On Becoming Human*. Cambridge: Cambridge University Press, 1981.

Tattersall, Ian, and Jeffrey H. Schwartz. "Evolution of the Genus *Homo*." *Annual Review of Earth and Planetary Sciences* 37 (2009) 67–93.

Teilhard de Chardin, Pierre. *The Phenomenon of Man*. Translated by Bernard Wall. New York: Harper & Row, 1975.

Tertullian. *The Apology*. Translated by S. Thelwall. In *Ante-Nicene Fathers*, edited by Alexander Roberts, James Donaldson, and A. Cleveland Coxe, vol. 3. Buffalo, NY: Christian Literature, 1885. Revised and edited for New Advent by Kevin Knight. http://www.newadvent.org/fathers/0301.htm.

Theissen, Gerd. *Biblical Faith: An Evolutionary Approach*. Translated by John Bowden. London: SCM, 1984.

———. *A Theory of Primitive Christian Religion*. London: SCM, 1999.

Toolan, David. *At Home in the Cosmos*. Maryknoll, NY: Orbis, 2001.

Torrey, E. Fuller. *Evolving Brains, Emerging Gods: Early Humans and the Origins of Religion*. New York: Columbia University Press, 2017.

Tuttle, R. Howard. "Human Evolution." *Encyclopedia Britannica*. 2021. https://www.britannica.com/science/human-evolutio/Refinements-in-tool-design.

Tyrrell, Kelly April. "Oldest Fossils Ever Found Show Life on Earth Began before 3.5 Billion Years Ago." University of Wisconsin–Madison, News, December 18, 2017. https://news.wisc.edu/oldest-fossils-found-show-life-began-before-3-5-billion-years-ago/.

Wallace, Mark I., and Smith, Theophus H., eds., *Curing Violence: Essays on René Girard*. Sonoma, CA: Polebridge, 1994.

Wilber, Ken. *Up from Eden: A Transpersonal View of Human Evolution*. Wheaton, IL: Quest, 1996.

Wildiers, N. Max. *The Theologian and His Universe: Theology and Cosmology from the Middle Ages to the Present*. New York: Seabury, 1982.

Williams, James G. *The Bible, Violence, and the Sacred: Liberation from the Myth of Sanctioned Violence*. Valley Forge, PA: Trinity, 1995.

Yong, Ed. "A Shocking Find in a Neanderthal Cave in France." *The Atlantic*, May 25, 2016. https://www.theatlantic.com/science/archive/2016/the-astonishing-age-of-a-neanthertal-construction-site/484070/.

Zeyl, Donald, and Barbara Sattler. "Plato's *Timaeus*." *Stanford Encyclopedia of Philosophy*. 2019. https://plato.stanford.edu/archives/sum2019/entries/plato-timaeus/.

Index

A

Abba, 5, 192–94, 216–17, 224, 226, 232, 235, 249–50
Abraham, 20–21, 29, 52, 183, 187, 190–91, 209
Abrams, Nancy, 113
abyss, 55, 99, 226, 231, 233
Acheulean Period, 167
âdâm, 71, 74
Adam, 5, 10, 16, 21, 29, 64, 67, 70–72, 227
Africa, 124, 126, 128, 181
  northern, 129
African Rift Valley, 124
African savannah, 125, 138
age
  critical reason, 22
  scientific, 7, 37, 40
  technological, 65
  of evolution, 265
agriculture, 118
Ahura Mazda, 186–87
Akhenaten, 187
  heretic Pharaoh, 186
Alber, Karl, 266
Alison, James, 213–14, 228–31

Almagest, 103
Ælohîm, 54
alpha, 196, 213, 259
Ambrose of Milan, 5, 218, 232, 249
American piety, 159
AMH (anatomically modern humans), 128, 130
Amorites, 209
ancestors, 94, 97, 99, 120, 124, 127–28, 130–31, 135, 137, 144–45, 181, 186, 246
  biblical, 30
  early, 142, 148, 166, 168–69, 183–84, 207, 213
  hominid, 246
  prehistoric, 18, 232
  primate, 135
ancient
  cosmologies, 3, 97, 99
  Israel, 66, 73, 194, 209
  symbols of divinity, 4, 180
animals
  kingdom of, 133, 243
  symbolic, 182
Annat, 180
Anselm of Canterbury, 5, 221

(Anselm of Canterbury *continued*)
   satisfaction theory of, 223
anthropic principle, 112–13
anthropology, 101
   evolutionary, 141
Antipas, 249
anti-Semitism
   modern, 216
apes, 117, 126, 144, 206
appease, 181–83, 209
appeasement
   sacrificial, 194
Aquinas, Thomas, 25, 96, 224
archaeological record, 125, 136
archaic communities, 213
Aristotle, 96, 101, 120, 243
   epistemology of, 103
   metaphysics of, 248
   philosophy, 96, 103–4
Artemis, 180
Asherah, 187–88
Assyria, 191
astronomy, 103, 106
Atapuerca Mountains, 136
atmosphere, 16, 83–85
atoms, 8, 14, 76–78, 89, 115, 239, 245, 253
atonement, 8, 212, 221, 223
Augustine of Hippo, 5, 29, 102, 123, 219–21, 249
Australia, 3, 8, 18, 116, 129, 134, 173
Australopithecus, 117, 125
autocatalytic, 87
autographs
   original, 48
autonomy, 58, 72, 190
awakening, 138, 165, 168, 197, 237, 247, 260

Axial Age, 118
Ayala, Francisco, 110

B

Ba'al, 187–88
Babylon, 54, 98, 191
Bacon, Roger, 104
bacteria, 84, 113
Bader, Christopher, 159
Bailie, Gil, 195
bârak, 61–62, 70
Barfield, Owen, 119
Barrow, John, 112
Barth, Karl, 25
Bartlett, Anthoy, 201, 212, 218–25, 232, 249
Basil the Great, 102, 262
basilea, 192–93, 217
Baudler, Georg, 24, 30, 143, 146–47, 168–69, 180, 183, 186
Baylor survey, 160
Being-in-Love, 255
Being-of-beings, 52, 55
belief
   in God, 1, 105, 159, 176
   structures of, 57, 215
   systems of, 18
believing, 28, 38, 76, 113, 148, 198
Bible, 3, 8, 21, 43–45, 47–49, 51–56, 60–61, 63–64, 66–67, 70–71, 73, 183, 189
biblical, 61, 66, 74, 123, 139
   traditional, 258
   truth, 21
   understanding, 62, 196, 228
Big Bang, 6, 26–27, 40, 80, 83–84, 115, 117, 170, 239, 256

cosmology of, 8, 197, 257–58
theory of, 82
biological change, 39–40, 73, 93
biology
   developmental, 4, 8
   evolutionary, 197
birth
   galactic, 245
Biser, Eugene, 146, 212
Blombos Cave, 131, 145
blood rites, 184
body size, 125, 149
Bohm, David, 108–9, 238–39
   implicate order, 109
Bohr, Niels, 237
bonobos, 125–26
Book of Revelation, 48, 175, 255
Bor, Daniel, 152
Born, Max, 237
brain, 4, 125, 127, 135–36, 140, 142–44, 148–61, 164–65, 167, 170
   evolving, 99, 213
   mystical, 165
   prehuman, 208
   protohuman, 167
   spiritual, 156–58, 160–61, 164–65
Braterman, Paul, 12
Brecht, Martin, 216
Brueggemann, Walter, 46, 49–51, 66, 74, 262
Bruniquel Cave, 18
Brunner, Emil, 21, 262
bull power, 186
Burkert, Walter, 147

C

Cain, 227–28
Calvin, John, 220
Calvinism, 220
capacities, 6, 13, 15, 130, 133–34, 136, 140, 143, 167–68, 171, 200, 205–6, 232
   human representational, 7, 229
   meaning-making, 213
carbon, 81, 89, 112, 119
caregivers, 134–35
   primary, 214
Catholic Church, 216, 224
Catholicism, 13, 36
cave paintings, 134, 147, 172
caves, 131, 136, 173
cells, 84–86, 88–91, 109, 114–15, 117, 149, 239
   first living, 10, 84
   first nucleated, 138
change
   adaptive, 255
   creative, 110
chaos, 46, 83, 97, 99, 210
   primordial, 99
Chapelle-aux-Saints, 136
Chardin, Pierre Teilhard de, 16, 247–48
chemistry, 12, 46, 87, 89, 109, 167, 198
chimpanzees, 133
Chouraqui, André, 57
Christ, 1, 3, 6, 9, 15, 19, 16, 25, 27, 31, 102, 104, 117–19, 121, 190, 194–95, 198, 216, 218–21, 224–26, 233, 255

(Christ *continued*)
  death of, 218, 220–21, 224, 225
  love of, 260
  passion of, 224
  resurrection of, 226
  sacrifice of, 220
Christian
  doctrine of creation, 8, 44
  cultural competence, 198
  eschatology, 259
  expectancy, 259
  experience, 59, 251
  faith, 1, 7, 95, 247
  God, 249
  imagination, 11, 225, 232
  inertia, 5, 247
  orthodoxy, 197, 215, 220
  religion, 179, 212
  revelation, 36, 124, 201, 212
  tradition, 10, 22, 30, 119, 235, 247
  West, 218
Christianity, 3–4, 10, 13–14, 41, 47, 116, 119, 218, 221, 224, 235–36, 239, 247–49, 257–58
church attendance, dwindling, 36
  life, 225
classical antiquity, 3, 100–101, 120
classical theism, 166
Claudius Ptolemy, 103
Clement of Alexandria, 102
Clottes, Jean, 173
clouds, 8, 78, 80, 104, 160, 168
  dust, 79
cognition, 133, 135, 142, 144, 149, 154–55, 171

growing, 198
higher, 169
human, 148, 187
symbolic, 131, 175
coherence, 3, 14, 87, 95–97
  inner, 16
  panoramic, 164
  preconceived theological, 252
  universal, 95
cohesion, 13, 32, 88, 211
community, 175, 183, 198, 208
  new, 183
  religious, 28
  scientific, 40
compassion, 158, 216, 246–47, 250–51
  abyssal, 218, 226, 231
  authentic, 201, 232
  bottomless, 249
  limitless, 226, 233
complexity and mystery, 4, 148, 153
conception, 15, 159, 163, 166, 197, 199, 211, 213, 216–18, 225–26, 236–37, 252, 254–55
  cosmological, 13
  outdated cosmic, 19
  sacrificial, 183
Confucius, 185
connectedness, 164, 166, 241
consciousness, 4, 119, 146–47, 150–51, 153, 155–56, 166–67, 182, 185–86, 197, 213–14, 231, 239–40, 242–43, 252–53
  higher, 122, 139
  holistic, 165
  modern, 147
  mystical-spiritual, 166
  of death, 226, 228, 233

INDEX

relational, 162–63, 167
religious, 197
self-reflective, 118, 121, 198, 242
unitive, 164
context, 31, 33, 47, 51, 66, 71–72, 95, 97, 162, 164, 166, 168, 17
communal, 3
cultural, 45, 140, 156
original, 100, 172, 180
controversy, 47, 123, 264
creation-evolution, 43
Copernican revolution, 105, 120, 236
Copernicus, Nicolaus, 38–39, 104
cosmic, 5, 11, 15, 19, 26, 28, 41, 64, 198, 255–57
absurdity, 258
awakening of love, 121, 242
calendar, 117, 199
Christ, 15, 42
drama, 137
ecology, 245
fine-tuning, 3, 110
history, 7–8, 11, 83, 92, 94, 115, 117–19, 167, 170, 196–97, 239, 242–43, 245
horizon, 92, 114–15
insideness, 14, 148
inwardness, 243, 245, 260
order, 101
organization, 245
perspective, 3
process, 20, 76, 108–9, 119, 148, 238, 241, 248
self-consciousness, 121, 141

story (new), 11, 14, 42, 46, 76–91, 107–8, 239, 245, 247, 253, 260
cosmology, 4, 8, 25, 28, 37, 39, 41, 96–102, 120–21, 259
evolutionary, 73, 197
modern, 25, 42, 48, 62, 247, 257–58
outdated, 13, 197
cosmos, 4–5, 10–11, 13–16, 25–27, 30, 44–45, 58–61, 63–68, 91–93, 96–97, 101, 105–7, 119– 20, 167, 235–37, 240–42, 245, 248, 258–60
cosmos and revelation, 2, 4–5, 8–10, 12, 14, 18–20, 124, 126, 128, 130, 132, 136, 138, 140, 144–252
Cowdell, Scott, 16
creation, 7, 4–16, 19, 22, 25–29, 34, 37–38, 40–41, 43–47, 54–67, 71–74, 100, 105–6, 119–20, 196–99, 234–35, 243, 247–48, 251–56, 258–60
accounts of, 28, 43, 45, 47, 50, 56, 62–63, 100
first, 53, 56, 61, 64, 74, 188
groaning of, 58
second, 51–52, 57, 187
story of, 51, 63, 72, 74, 190
creation-evolution debate, 6, 71, 74
creativity, 5, 40, 46, 58–59, 92, 121, 167
creator, 4–5, 8–9, 11, 14–15, 19–20, 40, 46–47, 58, 62, 66–67, 69–70, 74–75, 107, 189–90, 243–44

creatures, 47, 58–59, 61–62, 65, 123–24, 130, 133, 135, 138, 140, 223, 253, 256
   imitative, 155, 199, 232
   predetermined, 58
   sentient, 78
credibility, 8–9, 11, 22, 124
Cro-Magnons, 131, 138, 173, 246
crucified God, 31, 195–96
crucifixion, 30, 256
culture, 7–8, 97, 100, 116, 155, 180, 201–2, 205–6, 208–9, 211–12, 226, 228, 231–33
   and religion, 202, 211–12, 232–33
Cur Deus Homo, 222–24

D

dark matter, 115
Darwin, Charles, 11, 39, 110, 123, 237
Davies, Paul, 113
death, 10, 13, 60, 184–86, 194–95, 213, 216, 218, 220, 223–24, 226, 228–31, 233, 251–52, 255–56
deep-sea vents, 84, 197
deities, 35, 59, 101, 186–88, 190, 237
   destructive, 183
   distant, 67
   local, 187
   warring, 100
Delio, Ilia, 104, 109
Denisovans, 138
Descartes, René, 105, 239, 243
   dualism of, 105, 236–37

De Trinitate, 219
devil, 218–19, 221, 224
dialogue, 29–30, 56, 61, 64, 67, 72, 74, 180, 190, 257
Dicke, Robert, 82
dinosaurs, 85, 92
disorder, 46, 97, 144, 200
diversity, 5, 39, 58, 70, 72, 83, 100, 118, 253–54
   biological, 128
divine, 29, 34, 41, 54, 59, 96, 99, 102, 175, 177, 180
   actions, 15, 252, 260
   anger, 209
   bull, 180
   communication and love, 253
   identity, 30, 195
   image, 260
   origin of revelation, 20
   promise, 27, 30–31, 259
   revelation, 7, 14, 20, 35, 38, 44, 77, 240, 243
   self-disclosure, 21, 23, 24, 30
   self-revelation, 9, 22, 23, 41, 42, 50, 66, 74, 122
   speech acts, 56
   violence, 212, 220, 224–25, 233, 249
divinity, 4, 27, 32, 51, 110, 180–82
DNA, 39, 83, 87, 91, 109, 126, 152
Dobzhansky, Theodosius, 123, 126
Dodd, Harold Charles, 255
d'Olivet, Fabre, 54–55
Donald, Merlin, 133
Donaldson, Margaret, 161
dopamine, 143, 145

INDEX

Douglas-Klotz, Neil, 194
drama of cosmic history, 170
Dumouchel, Paul, 202
Dürr, Hans-Peter, 91
dust, cosmic, 79–80

E

early mammals, 117
earth, 10, 12–13, 37–40, 42–43, 53, 56–58, 76, 79–85, 99, 101, 103–5, 109–10, 112–13, 119–20, 197
   flat, 98–99, 120
earthquakes, 10, 58, 209
East Africa, 144
ecological niches
   new, 127
ecosystems, 87, 130, 253
Eden, 10, 68–69, 105, 116, 226
education, 7, 8, 159, 162
Edwards, Denis, 254, 263
Egypt, 3, 54, 99, 103, 123, 181, 186, 189, 191, 228
   mythology of, 99
Einstein, Albert, 107, 120, 237
   theory of, 238
El, 51, 148, 182–83, 187
El Elyon, 187–88
Eliade, Mircea, 175–79
Ellul, Jacques, 29, 43, 45, 49–58, 60–61, 63–65, 68–73
Elohim, 2, 24, 51–53, 55, 57–58, 64, 67–68, 71
Elohist tradition, 53
emergence, 12, 14, 82–84, 86, 131, 134, 143–45, 148, 168, 170, 206–7, 242, 245–46
   gradual, 119

human, 131
   of subjectivity, 170, 235, 239
English Reformation, 216
enlightenment, 185
Erickson, Millard, 48
eschatology, 257–59
Ethiopia, 17, 128
Eucharist, 220
Euclid, 244
eukaryotes, 86
Euler's law, 244
Europe, 13, 18, 29, 36, 127–29, 167, 173, 216, 222
Eve, 5, 29, 227
evolution, 5–6, 13, 15–16, 37, 39, 60, 86–87, 108–10, 121, 123, 170–71, 247, 254–56
   animal, 127
   biological, 7, 15, 79, 110, 190, 255–56
   cosmic, 15, 27, 95
   cultural, 5
   of culture and religion, 232
exile, 175, 187, 189, 191
Exodus, 128 187, 190, 228, 246
experience
   collective, 206
   symbolic, 207
   of love, 164
extinctions, 10, 116, 129, 170

F

fairy tales, 182
   ancient Semitic, 181
faith, 9, 13–14, 21, 31–32, 34–35, 41, 44, 95, 156, 158, 194, 197, 236
   intellectual, 221

Faraday, Michael, 13–14
fate, 15, 30, 180, 182, 209, 228
father, 35, 59, 182, 192–93, 200, 216, 224–25, 249, 251–53
father God, 193
fertility, 178, 180–82, 246
Fibonacci sequence, 244
findings
  scientific, 5, 8–9
Finlan, Stephen, 212
First Crusade, 222
Flinders Ranges, 129
floods, 10, 18, 181
food, 84–85, 105, 125–27, 139, 144, 172, 184, 213, 240, 247
food chain, 84, 129, 138, 147, 245
forebears
  ancient, 147, 197
forefathers, 147
  prehistoric, 148
forgiving victim, 5, 226, 231, 233
fossils, 10, 17, 126, 128
  hominid, 127
  modern human, 136
foundations
  of religion and culture, 208, 212
Fourier transform, 244
Fourth Gospel, 62
fragmentation, 108, 237–38
France, 18, 36, 128, 136
Frank, Adam, 107–8, 220
free
  energy, 84
  oxygen, 85, 117
  will, 64, 70, 200
freedom, 4–5, 24, 50–57, 60–61, 63, 65, 68–73, 185, 190–91, 226–27, 241, 244–45, 253–54, 256, 259
freedom and love, 15, 70, 226, 244, 253, 256
Funes, José G., 234
futurity, 27, 30

G

galactic supercluster, 115, 167
galaxies, 15, 20, 76–79, 81–82, 87, 93, 109, 112–13, 115, 150, 153, 167, 170
Galileo, 11, 103
Gallese, Vittorio, 205
Garrels, Scott R., 205, 214
Gautama Buddha, 186
Genesis, 4, 57–58, 61–62, 70, 98, 122, 162, 183, 187–90, 209, 227
  accounts, 11, 15, 60, 66, 73, 100, 124, 183
  authors of, 60
  text, 4, 19, 44, 124
genetic mutations, 143
  lethal, 10
Theissen, Gerd, 187, 192
Gibeon, 187
Gibson, Shimon, 188
Giegerich, Wolfgang, 184
Girard, René, 201–2, 204–8, 211–13, 215, 226, 230, 232
  anthropology of, 228
  mimetic theory of, 201, 204, 206, 212, 226–27
Gnuse, Robert, K., 187
Göbekli Tepe, 98
God, 1–2, 1–3, 5–16, 19–27, 29–35, 40–41, 43–74, 95–102,

104–7, 119–20, 122–24,
  156–63, 165–66, 185–87,
  189–99, 208–11, 216–36,
  246–51, 253–56
 biblical, 36, 190
 breath of, 249
 distant, 51, 159, 236
 omnipresent, 22
 personal, 15
 relational, 68
 transcendent, 21, 34
 triune, 252
God as Oneness, 51
God language, 212, 216–18, 232
God-of-the-future, 27
Golden Ratio, 244
Goldilocks enigma, 113
Gombe, 136
Goodall, Jane, 136
gospel, 3, 13, 46, 50, 217, 249, 251
grace, 3, 6, 24, 192
gravity, 37, 39, 77–80, 88, 108, 111, 113–14, 121, 157, 204
Greek mythology, 99, 185
Greenspan, Stanley, 134
Gregory the Great, 5, 220, 249
Grosseteste, Robert, 104

H

habitat, 67, 127
Hameroff, Stuart, R., 238
Hammurabi, 116
Hardy, Allister, 162
Haught, John, 22–23, 26–27, 31, 148, 170, 239, 241–42, 245, 247
Hawking, Stephen, 47, 111
Hay, David, 161–64
heart, 14, 16, 21, 30, 52, 67, 90, 165, 185, 253, 260
heavens, 33, 53, 62, 93, 98–99, 102, 106, 192, 195, 197, 217
Hebrew, 21, 43, 45, 53–54, 57, 60, 71–72, 100, 124, 181, 194
 ancient, 100
 Bible, 21, 24, 29, 33, 44
 text, 44–45, 48, 61, 67, 71, 73
 tradition, 30, 47, 56, 100
Heisenberg, Werner, 91, 237
heliocentric, 103
Hellenic period, 187
Henshilwood, Christopher, S., 131
Herod Antipas, 249
Hexaemeron, 102
hierophanies, 23, 175–79
Hillman, Ann, 165
historical situation, 7, 19, 176–77, 235
history, 20, 22–23, 26–31, 40–41, 44, 49–50, 65–67, 73–74, 83, 96–97, 105–6, 110–12, 119–23, 176–77, 195–96, 200–202, 230–32, 260
 archaeological, 130, 132
 authoritative, 33
 church, 215
 cultural, 222
 evolutionary, 122, 139, 207
 planetary, 84
 prehuman, 139
 redemptive, 29
 religious, 158, 178, 180, 186, 190
 sacred, 29
 theological, 50

(history *continued*)
  violent, 212
Hobbes, Thomas, 14
Hoffman, Joel M., 43, 45
holy, 32, 157, 175–77, 179, 218
Homo
  erectus, 3, 118, 126–29, 132, 136, 138, 142
  floresiensis, 132, 138
  habilis, 92, 117, 125–27, 129, 133
  heidelbergensis, 128–29, 138
  neanderthalensis, 132
  sapiens, 10, 17–18, 122, 126–28, 130–32, 134, 138, 167, 198, 240, 247
horizon, 8, 7, 11, 135–37, 146, 151, 168, 181, 260
  cosmological, 110
  ever-receding, 120
  transcendental, 182
Hosea, 188
Hubble, Edwin, 237
human
  conscience, 196
  consciousness, 7, 14, 19, 25, 142–71, 173, 176, 179, 206–7, 242, 245
  culture, 98, 204, 208, 211, 213–14
  intelligence, 143–44, 260
  emergence, 130–31
  life, 30, 47, 112, 257
  mind, 21, 41, 130–31, 135, 196–97, 244
  sacrifices, 184, 209–10
humans, 10, 21, 57, 62–65, 67, 69–70, 74, 98–101, 120–22, 168–69, 172–74, 229–30, 232, 242–43, 245
  archaic, 128
  dislodged, 105
  early, 133, 147, 168, 185
  universal mimesis, 204
hunter-gatherers, 173
hunters, 127, 129, 133, 143–44, 146, 148

I

identity, 30, 33, 36, 164, 195, 214–15, 224, 232, 249, 251
  collective, 175
  corporate, 226
  individual, 202, 212, 215, 232
illusion, 108–9, 184, 215
image, 19, 61–65, 70–72, 74, 101, 109, 122, 147, 159, 162, 251
image of God, 61, 63–64, 71–72, 74, 196, 225
imagination, 26, 28, 168, 174, 245
  creative, 37
  religious, 10, 197
imaginative remembering, 66, 73
imitation, 204–6, 214
imitative spiral, 202
immanence, 107, 230
  radical, 7, 231
incarnation, 8, 56, 119, 173, 218, 256
  paradox of, 179
  triumph of, 16
incoherence, 237
inerrant, 49

innovation, 7, 117
  evolutionary, 127
inquiry
  critical, 3, 235–36
inseparability
  of cosmos and Christ, 25
inspiration, 2, 4, 48
  plenary verbal, 48–49
intellectual integrity, 6, 14, 26,
  41, 43–44, 73
intelligence, 64, 70, 123, 167
  human, 145
  superior, 101, 120
interior horizon, 3, 135–36
interiority, 254
  conscious, 148
interventions, 254
  causal, 179
  miraculous, 119
  special divine, 5
intuition, 252
  ancient Hebrew, 139
Iran, 186
Iraq, 136
Isaac, 123, 187, 209
Isaiah, 100, 188, 249
ish, 71–72
isha, 71–72
Ishtar, 180
Isis, 180
Islam, 186, 248
Islamic monotheism, 181
Israel, 21–22, 24, 30, 33, 50–51,
  56–57, 66, 74, 101–2, 127–28,
  187–92, 227–28, 232–33,
  250–51
  early, 210
  history of, 187, 190
  memory of, 191, 246
  transcendent horizon of, 191,
    193
Italy, 127

J

James, Stephen R., 127
Janssen, Luke, 138
Jeans, James, 238
Jedi jaw, 17
Jeremiah, 54, 188
Jerusalem, 187–89, 251
  temple, 210
  (Bible) translation, 53
Jesus, 29–30, 35, 192–95, 197,
  199, 216–17, 223–24, 228–31,
  233, 235, 247, 249–52, 255–60
  Abba of, 192, 251
  crucifixion of, 259
  death of, 212, 226, 232–33,
    252
  resurrection of, 7, 8, 29–31,
    121, 218, 226, 229–33, 248,
    257–60
  understanding (of God) of,
    250
Jewish, 62, 189, 193
  religion, 24
  scriptures, 193, 209
  thought, 56, 59–60, 63, 68
Jews, 30, 32, 55–59, 63, 107, 189,
  193, 216, 222, 228
  massacred, 216
  postexilic, 189
Jonas, Hans, 241–42
Jordan, Pascual, 237
journey, 4, 16, 26, 29, 44, 76, 81,
  186, 191, 222, 228
  evolutionary, 169

(journey *continued*)
   human, 116, 122
   spiritual, 31
Judeo-Christian
   theism, 102
   tradition, 20, 23, 100, 156, 226
   worldview, 96
judgments
   apocalyptic, 191
Jupiter, 38, 81, 101

K

Kauffman, Stuart, 51
Kenya, 126
killing, 125, 147–48, 184, 213
kingdom, 29, 116, 123, 183, 192, 217, 250, 259
   peaceable, 250
kingship, 217
knowledge, 20, 23, 29, 34–35, 46, 50, 66, 68, 72, 93, 95–96, 98, 251
   culture-based, 45
   evolving, 96
   of death, 229, 231
   of God, 34, 96
   of science, 95

L

Lake Victoria, 124
land, 2, 17, 37, 56–57, 60, 63, 116, 138, 182, 188
language, 38, 45, 50, 52–53, 150, 152, 155, 172, 174, 176
Lao Tzu, 185
Latin, 53
Latin
   Church, 218
   Vulgate, 53
laws, 7, 36, 87, 102, 106, 111, 169, 189, 196, 216, 258–59
   physical, 106
   universal, 104
Leidenhag, Joanna, 240
Lewis, C. S., 31–32, 119
life, 5, 25–28, 32–33, 39, 59–60, 81–86, 90–91, 110–12, 118–19, 121, 172, 182–83, 185–86, 192, 194, 199–200, 229–31, 233, 240–41, 254–57
   bacterial, 197
   extraterrestrial, 15
   inner, 161
   intelligent, 150
   microbial, 84
   plant, 257
   religious, 191
   spiritual, 161
limbic system, 139
lipids, 87, 89–90
Logos, 255
Lohfink, Norbert, 187
lord, 222–24, 246, 250
Lord God, 146
los Huesos, 136
love, 3–4, 15–16, 24, 29, 50–57, 59–63, 65, 68–75, 158, 164–66, 194–95, 217–18, 225–26, 229–30, 250–51, 253–54, 260
   abyssal, 219
   creative, 6, 20
   crucified, 195, 199
   divine, 7, 65, 196, 228–29, 233, 255, 260
   dynamic, 253
   eternal, 254

feminine, 180
formative, 22, 41
freedom-granting, 245
gratuitous, triune, 260
sacrificial, 232
Lower Paleolithic, 167
Lücke, Ulrich, 137
Luke, 249–50
Luther, Martin, 5, 33, 216, 220, 225

M

Maccabees, 56
Mack, Burton L,. 201
Magisterium, 33
magnetic field, 82, 257
mammalian, 140, 200
mammals, 58, 85, 116, 134, 139, 149
mass (of matter), 40, 46, 56, 77–78, 80, 86, 106, 108, 111, 114, 188
material, 91, 100, 179, 216, 240
material reality, 237–38, 253
material world, 58, 62, 85, 91, 237, 240, 242, 247, 254
materialist, 237
materialist argument, 240
mathematics, 51, 104, 111, 234, 236, 244
Matthew's Gospel, 216
Maxwell, James, 237
McArthur, John, 97
McHenry, Henry, 128
meaningful wholes, 149
meaninglessness, 185–86
mechanism, 39, 105, 120, 144, 201, 207–8

evolutionary, 124
hidden mimetic, 208
scapegoat, 208, 211, 213, 227, 231
violence-channeling, 184
medieval, 31, 222
Christianity, 222
model (cosmic), 15
synthesis, 105
Meltzoff, Andrew, 205, 214
memory, 89, 109–10, 142, 144, 146–47, 152, 154–56, 158, 169
active, 140
earliest, 123
Mercury, 81, 101
Mesopotamia, 98, 209
Messiah, 228
metabolic processes, 88
metabolism, 91, 137, 241–42
metanarrative, 108
metaphors, 16, 32
incarnational, 244
metaphysics, 66, 74, 248
Micah, 188
Middle Ages, 2, 10, 25, 96, 103, 118, 220
Middle Stone Age, 265
Milky Way, 15, 18, 20, 78–79, 99, 111–12, 237, 241
mimesis, 203–6, 208, 211–12, 214, 232
conflict-prone, 232
institutionalized, 221
mediated, 205
mimetic, 201–6, 208
crisis, 203, 207, 211–12
rivalry, 204–6, 219, 222, 227, 232
violence, 207–8, 211

mind
- collective, 155–56
- contemplative, 162
- of Christ, 165
- sleep/wake, 242

mindless, 163, 239
Moab, 209
models, 60, 65, 94, 97, 100, 103, 115, 151, 155, 177, 180, 202–5, 214–15
- classical, 152
- legal, 220
- nonrivalistic, 231
- primitive, 129

modern
- humans, 92, 116, 118, 126, 128–30, 173
- science, 4, 8, 15, 17, 25, 28, 41, 46, 54, 93, 197

Mohammed, 32
molecules, 83, 87–91, 239, 253
Moltmann, Jürgen, 195–96, 248, 257
monotheism, 4, 25, 33, 51, 186–89, 191, 198
- emergence of, 187–88

moon, 37–38, 81, 93, 101, 106, 112, 165, 197, 200
Moore, Sebastian, 215
morality, 45, 101, 154, 157, 187, 196, 241
Moses, 21, 52, 188, 190–91, 246
Mount Horeb, 188
Mount Sinai, 246
movements
- evolutionary, 235
- stellar, 32
- tectonic, 244

murder, 193, 211

collective, 213
Muslims, 32
mysterium tremendum et fascinans, 32, 192, 246
mystery, 1, 4, 17–19, 21–23, 32–34, 41, 44, 130–31, 148, 150, 161–62, 164, 166, 177–78, 200–201
- numinous, 18
- of redemption, 44

mystical experiences, 161, 165–66
myths, 18, 29, 110, 148, 176, 179, 181, 211, 216

N

names for God, 2, 51
NASA, 241
national deity, 188
nations, 29, 54, 188
natural
- disasters, 10, 209
- order, 7, 10, 57, 92, 119, 124, 238
- sciences, 93, 96, 104–5, 111, 236, 239–41, 243, 257
- selection, 39, 110, 122, 134, 137, 172, 247
- theology, 5, 32, 34
- world, 4–5, 9–10, 17, 34, 40, 111, 176, 179, 235, 237–40, 243–44, 246, 253

nature
- creative, 259
- creative-process, 11
- evolutionary, 26
- human, 133, 195, 204, 225, 230

of religion, 210
quantum, 46
Neanderthals, 18, 127–30, 137–38
neocortex, 140
Neolithic Revolution, 116, 118
neurons, 4, 149, 151–53, 156, 164–65
neuroscience, 25, 151, 160–61, 176, 204
New Testament (NT), 21, 24, 30, 33, 35, 48, 50, 56, 59, 191, 195, 197
Newberg, Andrew, 156–58, 160–61, 164–65
Newton, Isaac, 106–8, 120, 236
Nietzsche, Friedrich, 185–86
Noah, 21
NT. *See* New Testament
nucleosynthesis, 80, 112, 150

O

objects
    astronomical, 106
obligations, 222, 224
Old Testament (OT), 46, 50, 55, 97, 99, 183, 188, 192, 209–10, 212, 227–28, 233
omega, 196, 248, 259
omnipotence, 59
omnipresence, 142
openness, 8, 72, 74, 190, 243
orbits, 40, 81, 103, 112, 120
order, 56, 60, 93, 97, 114, 121, 219, 223, 239, 244–45, 253
    ancient sacred, 227, 233
organization
    biological, 19, 167

origin
    cultural, 184
    evolutionary, 138, 153
    scene of, 208, 210
    supernatural, 148
    of sacrifice, 183
OT. *See* Old Testament
Otto, Rudolf, 32
Oughourlian, Jean-Michel, 204
oxygen, 81, 85–86, 88–89, 92, 112
    atmospheric, 85–86

P

Packer, J. I., 21
Palaver, Wolfgang, 201
Paleolithic, 25, 98, 137, 148
paleomammalian, 139–40
Panikkar, Raimundo, 252–53
Pannenberg, Wolfhart, 268
parables, 24, 172, 216
paradigm
    cultural, 162
    evolutionary, 226
paradise
    lost, 248
parents, 129, 155, 193, 203
participation, 141, 190, 256, 259
    selfless, 256
path
    conflicted, 105
    evolutionary, 208
    holistic, 158
    sacrificial, 184
patterns
    self-organizing, 119
    symmetrical, 86
    visual, 155

Paul, 14, 21, 35, 194, 251
peace, 2, 158, 163, 201, 207, 210, 212, 228, 232
Peacocke, Arthur, 12, 36–37, 257
pedagogy, 73–74
Peebles, Jim, 82
Penrose, Roger, 238
Pentateuch, 52, 189
Penzias, Arno, 82
people of God, 29
perceptions
  holistic-symbolic, 174
  human, 149
  personal, 170
  prescientific, 236
  social-relational, 164
perfection, 15, 57, 101, 168, 254
  primordial, 248
Persia, 187, 189
person, 21–23, 32, 34–35, 52, 54, 65, 68, 73, 154–55, 160, 176, 214, 218
perspective, 52, 58, 69, 71–72, 107–8, 160, 166, 170, 174, 180, 187
  biblical, 38, 68
  historical, 47
Peters, Ted, 16, 257–258
phases, 22, 41, 128
  creative, 131
  preanimate, 242
phenomena, 16, 24, 26, 34, 155, 161, 174, 176, 178, 240, 244
  astronomical, 103
  emergent, 61
  geological, 123
  religious, 123, 175–78
  social, 200

philosophy, 29, 34, 66, 74, 96, 100, 102–3, 151, 168–69
physics, 37, 39, 46, 49, 108, 110–11, 198, 204, 234–35, 238
  classical, 108–9, 151
  modern, 55, 83
Pilbara Craton, 10
Pivotal Age, *see* Axial Age
Planck, Max, 237
Planck length, 114–15
Planck scale, 238
planets, 5, 78–85, 88, 90, 92–93, 96, 100–101, 103, 110–12, 130, 132, 169–70, 240–41, 257
plants, 60, 85, 112, 123, 236
Plato, 96, 100–102, 120, 248
Pluto, 81
Polanyi, Michael, 241
Polkinghorne, John, 14, 118, 257
potency, 207, 247
potentialities
  contradictory, 201
potentiality, 55
power
  creative, 46, 59, 189, 245
  destructive, 182, 246
  divine, 51, 74
  eternal, 14, 19
  gravitational, 195
  ife-generating, 260
  life-threatening, 184
  transcendent, 194
prayer, 158, 162, 193
prebiotic
  stage of cosmic history, 242
predation, 10, 245
predators, 133, 183
Primack, Joel, 113–14
primates, 117, 138, 140

Prime Mover, 103, 120, 247
primordial couple, 245
principle, 15, 34, 95, 97, 99, 172, 174, 183–84, 190, 236, 240
  cultic, 220
  dialogical, 61, 190
  fundamental, 231
processes
  belief-generating, 155
  biochemical, 83
  biological, 110, 198
  creative, 40, 59, 190, 239, 245, 254
  emotional, 159
nuclear, 81
profane, 176–78, 263
prohibitions, 33, 210–11, 227, 246
prolepsis, 259
proleptic, 29
Prometheus, 185
promises, 23, 26–27, 31, 33, 46, 123, 194, 228, 233, 236, 258
  implicit, 199
proofs, 25, 33, 102, 155, 157, 258
  observational, 237
  scientific, 83
prophets, 20–21, 33, 46, 52, 188, 191, 194, 250
  classical, 188
Protestant Reformation, 13, 33, 36, 225, 232
protogalactic disks, 78–79
Ptolemaic cosmology, 105
Ptolemy's universe, 103
public revelation, 1, 31, 33, 40, 141

Q

quantum, 113, 114, 237–38
quarks, 77, 253
Qu'ran, 32

R

radiation, 77
  cosmic microwave background, 77, 82
  deadly cosmic, 257
radiation era, 77
Rahner, Karl, 41
Ramsey, Ian, 174
Rashdall, Ian, 223
realism
  critical, 104
reality
  absolute, 177
  biological, 230, 237
  constructed, 156
  experiential, 165
  historical, 177, 233
  human, 226, 230
  metaphysical, 203
  sacred, 176
  spiritual, 160
reason, 45, 101, 117, 157, 169, 200, 217, 222
  critical, 20, 22
  human, 51
redemption, 44, 158, 183, 219–20, 224, 256, 260
  mystery of, 44
Rees, Martin, 111
Reformation, 2, 12, 97, 118, 249
relationship
  dialogical, 199

(relationship *continued*)
   enfolded, 108, 238
   fractured, 99
   human-divine, 190
   intimate, 193
   of love, 69–70, 72, 74
   religion, 12, 18, 29–30, 36–38, 119–20, 153–55, 157–59, 174–76, 179–80, 187–88, 201–2, 208–12, 231–33
     archaic, 201, 208–10, 227
     folk, 221
     historical, 18, 119
     institutional, 166
Renaissance, 31, 105, 120
resurrection, 7–8, 29, 31, 121, 218, 229–31, 233, 248, 256–57, 259–60
revelation, 1–2, 4–5, 2–260
   biblical, 21, 29, 31, 56, 62, 73
   personal, 2, 31–33
   progressive, 73
   public, 31, 33, 40, 141
   in Christ, 198
   in history, 1, 20, 22–23, 27–28, 30–31
   of divine love, 7
   of God, 4, 44, 50, 194, 226, 235
   science as, 38
ritual killing, 210
rituals, 18–19, 166, 172, 190, 246
   cultic-sacrificial, 147
rivalry, 164, 205, 214, 217, 227
rivals, 203, 207, 214
Russell, Heidi Ann, 96
Russell, Robert John, 256–259
Rovelli, Carlo, 245

S

Sacred (the), 51, 156, 176
sacred
   awe, 213
   status, 222
sacrifice, 5, 181–84, 190, 200–233, 246
   voluntary, 209
sacrificial, 185
Sagan, Carl, 94, 113, 116
scapegoating, 201
scapegoats, 5, 184, 201, 208, 216, 224, 232
Schroeder, Gerald, L., 53, 91
Schwager, Raymund, 184, 201, 226
science
   discoveries of, 20, 43, 95, 248
   empirical, 103
   evolutionary, 14
   experimental, 104
   modern materialist, 148
scientific evidence, 5, 9, 39, 73, 91, 123
secular, 26, 163, 176, 215
self, 163, 215
   assertion, 29, 185, 194
   communication, 22, 64, 253
   gift, 22, 27, 251
   organization, 84, 86–87
   revelation, 9, 22, 27, 41–42, 50, 66, 74, 122
self-disclosure, *see* divine
self-revelation, *see* divine
sinful, 21, 196, 204, 226, 230
singularity, 111, 255
sinners, 196, 223, 225

sins, 183, 195, 204, 209, 220, 223–25, 227, 230, 251
size
   human, 113
   physical world, 113
   scales, 113–15
   spectrum, 113
Smith, John, 187
Smithsonian, 172
social cohesion, 191, 211
Sollereder, Bethany, 16
Son, 59, 182–83, 216, 220, 224, 249, 251–53, 255
son of man, 192
source, 10, 13, 20, 46, 48, 69, 101, 105, 214, 224
   extrabiblical, 189
   scientific, 138
Southgate, Christopher, 12, 16, 58
space, 4, 58–59, 77, 79–80, 82, 84, 90, 92, 94, 106–7, 114, 120–21, 237–40
space-time, 7, 46, 78, 92, 108, 113–14, 121, 237–39, 254
species, 28, 60, 62, 65, 116, 123, 125–26, 128–32, 135, 141, 201
   hominid, 132
   human, 9, 62, 126, 129, 138
   humanlike, 125
   mammal, 128
spirituality, 142, 151, 154, 156, 159, 161–63
   of children, 4, 142, 161, 164
star furnaces, 118
stars, 76, 78–81, 83, 87, 93–94, 99–100, 103, 110–11, 113, 118–19, 150, 168, 170
   first-generation, 112

neutron, 110
stellar furnaces, 84
substitution
   penal, 218–19, 224–25
sun, 15, 37–38, 40, 79–82, 85–86, 94, 99, 101, 103, 110, 112–13, 115, 119–20
supernova explosions, 80–84
Swaminathan, Nikhil, 149
Swimme, Brian, 81, 269
symbol
   formation of, 171
   of fertility, 181
   intelligence, 208
   thinking, 134, 166, 205
   central in Christianity, 195
symmetry, 46, 78
synapse, 152
systems, 11, 29, 46, 66, 86–87, 149, 151, 153, 179, 193, 223
   self-organizing, 87, 149, 154

T

Tanner, Nancy, 137
Tanzania, 3, 124, 126, 210
technology, 6, 8, 41, 179
   blade-making, 129
   stone tool, 129
Terra Amata site, 128
tetragrammaton, 52, 187
Theissen, Gerd, 179, 191, 193
theistic evolutionists, 5
theophanies, 23, 246
theory, 26, 39–40, 63, 107, 120, 179, 183–84, 204, 218, 243
   quantum, 108, 237
   of creation, 63
   of cultural origins, 184

(theory *continued*)
  of knowledge, 3, 95–96
Tipler, Frank, 112
tohu wabohu, 55, 60, 63
toolmaking, 129, 131–133
tools, 91, 106, 114, 133, 167, 184, 211
  intellectual, 37
  primitive, 133
torah, 72, 193, 227–28, 233, 246
traditions
  ancient sacral, 226
  mystical, 166
  old oral, 189
  oral, 33, 175
  philosophical, 162
  religious, 19, 181, 253
  theological, 70
transcendence, 7, 105, 117, 177, 201
transcendent, 22–23, 32, 98, 169, 176, 178–79, 196, 219
  dimension, 10, 24, 148, 169, 173, 181, 246
  horizon, 137, 191, 194
  reality, 23, 29–30, 32, 193
transformation
  redemptive, 255, 258
transgression, 146, 227
  human, 57
translations, 43, 48, 54, 57, 64, 184, 193, 265
  literal, 57
travail, 2, 78
trinitarian love, 9, 217, 252, 254, 259
Trinity, 5, 235, 248, 252–55, 259
  in creation and evolution, 235
truth
  ultimate, 156
  public claims, 236
  finding, 95–96
tsunamis, 58, 245

U

Uganda, 210
Ugaritic
  Ba'al, 180
  tradition, 182
ultimate, 46, 105, 196, 256
  convergence, 16
  destiny, 248, 256, 259
  reality, 1, 23–24, 156–57, 159, 177
  union, 30, 248
unfolding, 6, 26, 29, 93, 107, 118–19, 121, 123–24, 253, 256
United States, 3, 39, 159
unity, 27, 59, 71, 74, 105, 121, 163–64, 168, 194, 207
universe, 2–4, 7–8, 13, 26–28, 36–38, 46–47, 61–62, 76–84, 91–95, 97–121, 138, 150, 165–68, 170, 196–99, 236–42, 245, 257–60
  clockwork, 236
  contingent, 259
  evolutionary, 48, 62, 167, 198
  expanding, 78, 81
  incomplete, 256
  intelligible, 38
  moral, 255
  observable, 26, 109, 114
  unfinished, 26
Upper Paleolithic, 129, 131, 173, 181
ur-violence, 213

V

Vanderburg, Willem H., 45
vengeance, 211, 220, 223, 227, 232
Venus, 81, 101
victim, 5, 183, 190, 207–8, 210–13, 230
   dead, 208
   expelled, 210
   innocent, 224
   risen, 230
victimage, 208, 214, 232
   mechanism of, 211, 214–15, 232
violence, 180, 183–84, 191, 194, 200–202, 207–8, 210–12, 217–18, 224, 227, 230
   blind elemental, 180
   chaotic, 210
   collective, 183, 207, 216
   religious, 184, 211, 249
   sacred, 211, 227, 233
   unstoppable, 206, 208
Virgo cluster, 79
vision, 15–16, 19, 33, 40–41, 46, 71, 74, 186, 192
   cosmological-theological, 13
voice, 3, 12, 66, 139, 191–92, 203, 246
   prophetic, 66, 73, 249

W

Wallace, Mark, 201, 265–66
West Semitic religions, 187
Western
   Christianity, 216–17, 226, 230
   civilization, 4, 104, 248
   culture, 5, 13, 60, 95–96, 111, 193, 236–37, 243
   Europe, 13, 36, 173
   theological canon, 221
wholeness, 6, 14, 19, 105, 108, 158, 164, 235–36, 238, 260
   supreme, 141
   unitive, 254
   world's, 169
Wilber, Ken, 105
Wildiers, N. Max, 101
Williams, James, 201
Wilson, Robert, 82
wisdom, 9, 18, 54, 91–92, 182, 185, 234
   all-encompassing, 46
   path of, 185–86
   God of, 53
witness, apostolic, 29
woman, 57, 62, 64, 70–72, 74, 193, 204
women, 62, 116, 159, 212
world, 14–20, 22–23, 31–33, 41–42, 49–50, 58–59, 94–97, 99–101, 104–5, 110–11, 156–58, 160, 162–64, 171–72, 174–77, 179, 195–200, 217–18, 236–37, 245
   biblical, 22
   new picture of, 8, 235–36, 254
worldviews, 31–32, 120, 175
wrath, 181, 183, 209, 225, 255

Y

Yahwism, 188
Yahwist tradition, 53
Yahwistic temple, 189

YHWH, 24, 30, 49, 51–53, 66–70, 100, 187–91, 226, 233
   caring nature of, 192
   commonwealth of, 192

Zone
   habitable, 110, 112, 257
Zoroastrianism, 117, 187

Z

Zambia, 128
Zarathustra, 186–87

www.ingramcontent.com/pod-product-compliance
Lightning Source LLC
Chambersburg PA
CBHW071234230426
43668CB00011B/1429